E-Book inside

Please register on:

https://www.pmcmedia.com/e-book-download

and use this publication as an e-book for free.

The download code for this book is: **6ZPC79**

FABIAN HANSMANN | WOLFGANG NEMETZ | RICHARD SPOORS

KEEPING TRACK OF TRACK GEOMETRY

BASICS – ERROR DETECTION – CORRECTION – QUALITY

A comparative overview of the DACH countries and the UK

PMC

Bibliographic information published by the Deutsche Nationalbibliothek:
The German National Library catalogues this publication in the German National Bibliography;
detailed bibliographic information can be found on http://dnb.de

Publishing House:	PMC Media House GmbH Werkstättenstraße 18 D-51379 Leverkusen
Office Hamburg:	Frankenstraße 29 D-20097 Hamburg Phone: +49 (0) 40 228679 506 Fax: +49 (0) 40 228679 503 Web: www.pmcmedia.com; E-mail: office@pmcmedia.com
Managing Director:	Silvia Goronzy
Publisher/COO:	Detlev K. Suchanek
Editorial Office:	Dr Bettina Guiot
Distribution/Book Service:	Sabine Braun
Cover Design:	Pierpaolo Cuozzo (TZ-Verlag & Print GmbH, Roßdorf)
Typesetting and Printing:	TZ-Verlag & Print GmbH, Roßdorf

© 2021 PMC Media House GmbH

1st Edition 2021

ISBN 978-3-96245-225-4

A publication by

International Publishing

Preface

Dear colleagues!

The title "Keeping Track of Track Geometry" is more than a pun chosen deliberately. This book aims to provide a basic understanding of the track system and its maintenance. The track geometry of a railway is the fundamental factor of its quality. Knowledge about track components and their basic interactions is key to the understanding, assessment and management of the role of quality in the system. Only the best possible quality will ensure maximum sustainability and form the cornerstone for the success of the system.

The six chapters of this book not only provide a well-founded introduction to the basic aspects of railway track and track construction, but also give a comparison of the engineering principles adopted in the DACH countries (Germany, Austria and Switzerland) and the United Kingdom. This comparison provides an interesting insight into the practical implementation of those technical principles and illustrates common features as well as differences. From the legal provisions to the basics of the permanent way as well as measurement methods, the book describes the key concepts that are required for the correction of track geometry on plain line track. One of our main objectives was to provide unbiased product and technological information; therefore, we have deliberately left out technology-specific details in some places.

After months of intense, exciting and sometimes also tense work on the manuscript, corrections, rewriting and deleting, it is a special moment to hold the finished book in our hands – a moment we would like to dedicate to those people who have accompanied and supported us through this time. In particular, for the first edition in the German language, we are thinking of Dr Matthias Landgraf and Werner Schachner. Throughout the project, we were able to count on the unquestionable support of our employers Plasser & Theurer and PMC Rail International Academy. Therefore, on behalf of all the assistance we have been given, we would like to express our sincere gratitude to Johannes Max-Theurer, Johann Dumser and Antonio Intini. We hope you enjoy reading our book and look forward to your feedback.

This copy is the second edition, which was initially translated from the first edition into English. The core text and layout of the book remains unchanged. What is new is the introduction of information concerning the principal mainland railway of the United Kingdom, formerly known as British Rail. Today, the former British Rail railway infrastructure is owned and managed by Network Rail, an arm's length public body of the UK Government. We hope that this second edition will provide the reader with an interesting contrast between track technical matters both in the DACH countries and the UK.

We would like to thank, Deutsche Bahn, Network Rail, OEBB and SBB for access to their track technical standards and permission to use specific data from them. Additional sources of information and more general background references are included in the Bibliography.

Fabian Hansmann | Wolfgang Nemetz | Richard Spoors

Notes on use:
In addition to legal provisions, technical principles are discussed using extracts from standards or internal railway rules and regulations. Any given limit values are examples and have been taken from the rules and standards quoted. These are subject to change. Therefore, the issue date of the source has to be considered. This book helps to depict interactions between individual topics but cannot and should not be a replacement for studying the respective rule or standard.
Unless quoted otherwise, the authors have the property rights of the figures used in the book. We would like to thank MATISA, Plasser & Theurer, PMC Rail International Academy, ROBEL and Trimble for kindly providing their photos and further documents.

Contents

Contents

1 Legal and technical rules and standards

Fabian Hansmann and Richard Spoors

1.1 Key issues

This chapter provides an overview of the legal and technical rules and standards in Germany (DB Netz), Austria (ÖBB Infrastruktur) Switzerland (SBB), (also known as the DACH countries) and the United Kingdom (Network Rail). The chapter will provide answers to the following questions:

- Which legal provisions are relevant for the maintenance of the railway infrastructure in these countries?
- What cross-border technical regulations exist and to what extent are they legally binding?
- What content is regulated by TSI Infrastructure?
- What differences are there between the ISO and CEN standards?
- What technical rules and standards apply on the networks of DB, ÖBB, SBB and NR?

1.2 The railway and its rules and standards

When the steam powered railway started operation in the early 19th century, individual countries developed numerous rules and regulations to keep the use of this "dangerous" new means of transport under control. By 1883 in the UK there were around 130 private railway companies with their own Rule Books, primarily focused on the safe operation of trains.

In 1879 the German Imperial Court, as a consequence of a claim for damages, defined a railway company as follows:

"A company, focused on the repeated moving of persons or things over not insignificant spatial distances on a metal foundation, which due to its consistency, construction and smoothness is designed to enable the transport of large amounts of weight and the achievement of a relatively significant speed of the transport movement, and as a result of this characteristic feature combined with the natural forces (steam, electricity, animal or human muscular activity, in the case of an incline of the railway also the own weight of the transport vessel and its load, etc.) used to generate the movement that is capable of generating a relatively powerful effect (depending on circumstances only useful by object, or also causing the destruction of human life or injury to human health) when being operated." [1]

Such a detailed definition of a railway company may be exaggerated but back then was probably due to the general scepticism towards the new means of transport.

But what does the term "railway" entail?

A clear definition of the term railway in the German language is not irrelevant. Its fuzziness in everyday use has a significant impact on legal and technical framework conditions. Here, two aspects have to be taken into consideration:

1. In contrast to other means of transport, such as the car, the railway in its original form constituted the construction and operation of the asset. Only as late as 2001, the European Union would pave the way for free access to the infrastructure by separating this original concept [2]. Today a clear distinction is made between railway infrastructure managers and railway undertakings.

2. When applying various rules and standards, attention needs to be paid to which parts of the system these apply. Some countries have different guidelines for main line tracks, trams, underground railways etc.

The numerous acts of Parliament, laws, regulations, guidelines and work instructions of all kinds make it difficult to keep an overview. The following section will aim to provide the basic correlations between the legal and technical rules and standards in the four countries (Fig. 1-1). It will concentrate on the issues relating to the rail infrastructure.

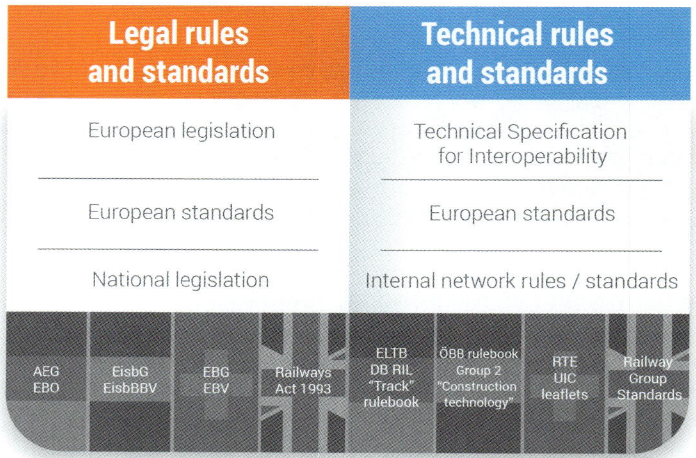

Fig. 1-1: Legal and technical rules and standards

1.3 From the technical entity to interoperability

The first railway networks were formed as self-contained networks, and their rules and standards were heavily dominated by factory standards. [3, p. 20]

A standard is a "document drafted by consensus and approved by an acknowledged institution. It specifies rules, guidelines or properties for general or recurrent use, pertaining to activities or products thereof. Its ultimate goal is to achieve the highest possible degree of order in a specific context." [4, p. 25]

With the rail connection between Aachen in Germany and Liège in Belgium the first cross-border railway line was opened in 1843. No matter whether the railway companies were under public or private management, it became apparent back then that it would be necessary to harmonise the technical framework conditions. The low flexibility of the gauge-bound system made it more difficult to meet the economical requirement of travelling over cross-border networks, which were very different from each other. In contrast, the military had a strategic interest in restricting and controlling cross-border traffic as much as it could. This can still be seen clearly in the different selection of track gauges throughout Europe (Fig. 1-2).

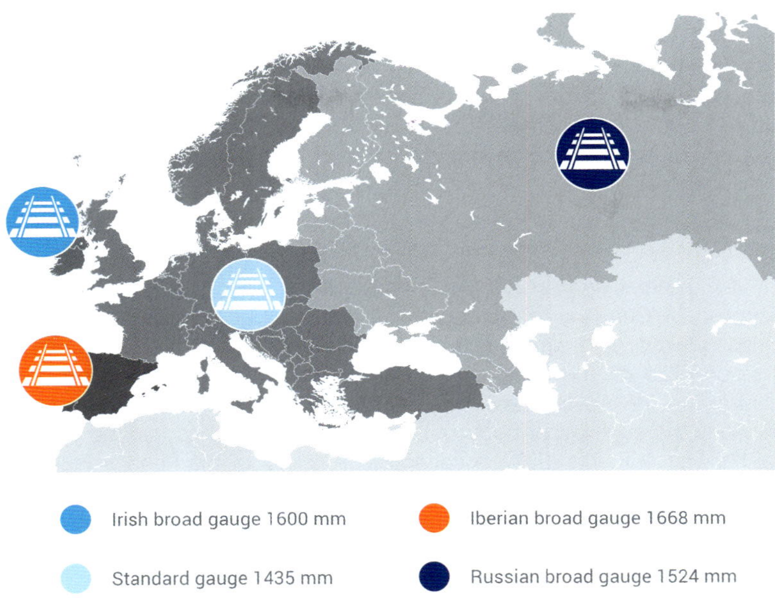

Fig. 1-2: Different track gauges as an example of the strategic role of the railway

1.3.1 Technical unity

Despite concerns from the military, Germany, France, Italy, Austria, Hungary and Switzerland laid the foundation for the **Technical unity of the railway sector** (Conférence internationale pour l'unité technique des chemins de fer) between 1882 and 1886. This is a state treaty that for the first time in the history of the railways defined essential parameters such as gauge, axle base, couplings etc. for cross-border traffic. Surprisingly, in 1938 this agreement gained validity throughout Europe, a long time before the European Union was formed and the term 'interoperability' became a hot topic. The treaty was not annulled until the end of the 20th century. Individual passages can still be found in the current European rules and standards. In the UK one of the early Acts of Parliament that concerned itself with the track was enacted in 1846 [5]. This mandated standard gauges of 4 ft 8 1⁄2 in (1,435 mm) for Great Britain, and 5 ft 3 in (1,600 mm) for Ireland. It also forbade the Great Western Railway laying any more of its broad gauge track of 7 ft 0 ¼ in (2,140 mm).

In Europe, the **International Railway Union UIC** (Union Internationale des Chemins de fer) was founded on 17th October 1922 in Paris. 51 founding members from 29 countries committed to standardisation and improvement of the construction and operation of railway systems. These days the organisation no longer consists exclusively of railway companies. The 200 members also include companies and suppliers from all over the world. UIC enjoys an outstanding reputation all over the world as a notable trade association for the railway sector. Although its significance has been reduced considerably due to the European railway packages and directives, it still publishes numerous UIC leaflets on different subjects (e.g. the UIC codex 720 on CWR track installation and maintenance). Of particular relevance for track construction are the leaflets of groups 71 and 72. Despite being widely accepted, they are only legally binding if there is a contractual agreement in place. It is planned for the leaflets to be gradually transferred to or replaced by the new platform "International Railway Systems" (IRS).

1.3.2 Legal foundations

Owing to the significant socio-political standing of railways, comprehensive legislation governs the framework conditions for the operation, construction and contract awards of public railway lines. Figure 1-1 shows a selection of the legislation that is currently in force. The details of the legislation are of less importance in this book than their impact on the follow-on rules and standards of individual railways.

1.3.2.1 Germany

In Germany the General Railway Act (AEG) provides the basic legal area for railways in the Federal Republic of Germany. [6] It applies to railways but excludes, for example, trams and mountain railways (and similar railways). Article 2(7) declares the railway infrastructure managers responsible for the construction, safe operation and maintenance of the railway lines. In contrast to the Austrian law, the German federal government, which according to AEG Article 5a is the supervisory body of the federal railways, has handed over this duty to the Federal Railway Authority (EBA) according to Article 3 of the Federal Railway Transport Administrative Law (BEVVG). [7] The EBA is an independent federal body which in Germany has the role of a supervisory, approvals and safety authority for railways and railway undertakings. As a public body, the EBA is responsible for the federal railway infrastructure, amongst other things.

According to Article 3(1) of BEVVG, the EBA fulfils the following duties:
1. Planning permission for operating facilities of the federal railways
2. Supervision of the railways
3. Construction supervision for operating facilities of the federal railways
4. Granting and revocation of operating licenses
5. Exercise of sovereign powers and of supervisory and participatory rights according to other laws and regulations
6. Preparation and execution of agreements as per Article 9 of the Federal Railways Improvements Act
7. (no longer applicable)
8. Authorisation of public funds for the promotion of railway traffic and the promotion of the combination of railway traffic with other types of transport.

In addition to the AEG, the Railway Construction and Operations Act (EBO) [8] is also relevant for the maintenance of the federal railways (public rail infrastructure managers). It consists of the following sections:
1. General
2. Railway installations
3. Vehicles
4. Railway operation
5. Personnel
6. Safety and order in the areas of railway installations
7. Conclusions
8. Appendices

In Germany, the EBA issues the Railway-specific List of Technical Construction Regulations (ELTB). ELTB defines those rules "that for the interpretation of Article 2(1) of EBO – 'Requirements of safety and order' are to be consulted regularly." [9, p. 1]

This list applies to railways owned by the federal government. For railways not owned by the government, the rules and standards of the Association of German Transport Undertakings (VDV) apply. [10]

1.3.2.2 Austria

The Austrian Railways Act (EisbG) captures all public railways (main lines, branch lines and trams) as well as private feeder lines and material trains. Which body is responsible for the individual railway (federal ministry or regional authority) depends on its classification into main line or branch line railway. [11]

The federal ministry of transport stipulates the basic requirements for safety, order and needs of the railway in regulations which also apply to the construction and maintenance of the infrastructure (Article 19(5) EisbG).

Every railway infrastructure manager requires a safety permit (Art. 38b EisbG) for operating main lines or associated branch lines; this can be issued by the ministry for a duration of up to five years. The introduction of a certified quality and safety management system is required for the permit to be issued. This management system monitors the compliance with the requirements of the Technical Specifications for Interoperability (TSI) and other national and international regulations in order to ensure the safe operation and safe maintenance of the railway infrastructure taking into consideration the state of the art. The Austrian Ministry for Transport, Innovation and Technology (BMVIT) can stipulate in regulations which requirements are necessary for meeting the state of the art (Art. 19(4) EisbG). Furthermore, it can publish schedules relating to regulations; these schedules detail provisions or other technical regulations concerning the basic state of the art (Art. 19(6) EisbG).

Art. 9b EisbG refers to the state of the art as follows:

"The state of the art for the purposes of this Federal Act is the development status based on relevant research findings of progressive technological procedures, facilities, methods of construction and operation, the functionality of which has been tried and tested. Determining the state of the art shall in particular be based on similar procedures, facilities, methods of construction and operation and shall take into consideration the proportionality between the effort of the required technical measures envisaged for the particular mode of operation and the benefits achieved for the interests to be protected."

Standards reflect the state of the art and the acknowledged rules of technology. [4] They, together with the internal ÖBB rules and standards, are considered as the tried and tested basis for managing the safe operation of railways on the ÖBB network. Put simply, standards form the basis for the technical policy, and the national rules and standards are their detailed interpretation. Standards may very well be legally binding. They become legally binding if laws and regulations refer to sections of a standard. For example, section 4.2.8.2 of TSI Infrastructure [12] refers to limit values in EN 13848-5 [13] and thus keeps these legally binding. However, not the whole standard is legally binding.

Apart from basic aspects of operational management, the Austrian Railway Construction and Operations Regulation (EisbBBV) [14] also defines, amongst others, terms such as track gauge, longitudinal gradient and clearance gauge and thus provides a description of essential components of the railway infrastructure. Among these is also a definition for maintenance in Art. 8(1): "The maintenance of the operating facilities and railway vehicles comprises servicing, inspections and repairs; it must as a minimum extend to all those areas, the condition of which can have an impact on operational safety and availability."

1.3.2.3 Switzerland

According to Article 1, the Swiss Railway Act (EBG) applies to the whole railway infrastructure on which licensed transport of passengers takes place or which is open for network access. [15] Therefore, this also includes trams and underground railways.

Although Switzerland is currently not a member of the European Union, Article 23a (7a) of the Swiss Railway Act states:

"Standard gauge railways must according to the provisions of this clause meet the technical and operational requirements for the safe and continuous train traffic in the European railway system (interoperability)." [15]

In addition to a valid license issued by the Swiss Federal Council, the European Safety Certificate is also required in Switzerland for managing the railway infrastructure. This certificate is issued by the Federal Transport Office (BAV) which acts as the supervisory body for all public transport and major parts of freight transport in Switzerland.

Art. 2 of the Regulation on Railway Construction and Operation (EBV) [16] clearly emphasises the importance of safe operation and appropriate maintenance. In principle, technical standards are to be drawn on and internal operating regulations adhered to in order to achieve these. In the Implementing Provisions of the Railway Regulation (AB-EBV) the BAV specifies which technical standards define the basic requirements of the system. [17, Implementing Provision 2.3]

These are as follows:

1. European Technical Specifications for Interoperability (TSI) as per EU Directive 2008/57/EC
2. CEN standards on railway applications
3. CENELEC standards on railway applications
4. ETSI standards (European Telecommunications Standards Institute)
5. Railway Technical Rules and Standards (Regelwerk Technik Eisenbahn RTE)
6. Guidelines of the Confederate Coordination Commission for Occupational Health and Safety (EKAS)
7. UIC leaflets
8. Agreement governing the exchange and use of coaches in international traffic (RIC)
9. Agreement governing the exchange and use of freight wagons in international traffic (RIV)
10. Recommendations of ITU (International Telecommunication Union) for transmission technology (ITU-T, analogue and digital)
11. SN 521 500 / SIA 500 "Barrier-free constructions"

1.3.2.4 United Kingdom

Early private railways were promoted by entrepreneurs who required an Act of Parliament to acquire the land on which their railway was to be constructed, the most well-known of which is the Stockton and Darlington Railway of 1825.

The Railway Regulation Act of 1840 [18] created a railways department within a Government department, the Board of Trade, through which the first railway inspectors were created. From now on, no new railway line could be opened without notification to the Board. These new inspectors also commenced the role of accident inspection and reporting. From 1842 a new act gave them authority to give direction to a railway if its safety system was poor. These were the first steps in the formation of Her Majesty's Railway Inspectorate (HMRI) [19] which continued as an independent government body overseeing railway safety until 1990 when it came under the Health and Safety Executive. In 2006 it moved again to the Office of Rail Regulation [20].

Legislation that defines specific engineering standards for the permanent way has only been introduced since 1996, with the onset of the privatisation of Britain's main line railways and, in more recent years, the creation of European Directives. For the first 75 years of railways, track standards (except gauge) were set by each private company's chief civil engineer. Commencing in 1901, the Railway Engineers' Association (REA) was formed by the chief civil engineers of the largest railway companies [21]. Its purpose was to discuss and agree common standards. The REA continued until the UK mainland railways were nationalised in 1948, when it changed its title to Civil Engineering Committee under the then Railway Executive. The Civil Engineering Committee, formed by the Chief Civil Engineers of the six railway regions, met, discussed, agreed and published Guidance Notes and Handbooks for use by Regional and District Engineers and their staff. This practice continued until 1993, when British Rail restructured its organisation in preparation for privatisation. A new Railway Group Standards unit was established, the purpose of which was to create and publish a suite of high level standards across the principal functions of engineering. The legal requirement for all entities involved in providing a mainline railway service to comply with Railway Group Standards commenced on April 1st 1994 with the enactment of the 1993 Railways Act and the creation of a new Track Authority called Railtrack. This legislation created a totally new railway structure in the UK by separating the management of the infrastructure from the operation of trains. Railtrack, the predecessor to Network Rail, required a Railway Safety Case and a Licence from the new Rail Regulator in order to discharge its responsibilities to manage the infrastructure and grant access to its tracks and stations for the train operating companies. Over the next nine years the UK rail industry went through a number of structural changes. In 2003 the Government established the Railway Safety and Standards Board, which is an independent company owned by its stakeholders who are the principal players in the UK rail industry [22]. Its purpose is defined as "to lead and facilitate the rail industry's work to achieve continuous improvement in the health and safety performance of the railways in Great Britain." It is within this role that it owns and manages Railway Group Standards which includes the UK railway Rule Book.

The Rule Book is applicable to all staff and has been developed from over nearly 200 years of history and lessons learned from operations and accidents. The early 'rules' were Bye-Laws, a principle taken from the development of canals in the late 1700's. Early Rule Books published by railway companies were necessary as they defined matters of behaviour, trespass and how trains could be operated safely by observing lineside signals given by members of staff, before the introduction of mechanical signaling. In 1842 the Railway Clearing House was established with the role of facilitating the through booking and movement of passengers and goods. However, it also played a role in the harmonisation of operating standards across the growing number of railway companies when, in 1876, it published "Rules and regulations for working railways". By 1883 nearly all the 130 private railway companies were being operated to a common set of rules. The next big change came in 1923 when the private railway companies were amalgamated into the big four, meaning just four rule books instead of 130!

The nationalisation of the railways of Great Britain in 1948 signalled the creation of a single British Railway Rule Book which was published in 1950 and forms the basis of the document we have today, although the format has changed as it has now been divided into sections, proportional to the groups of staff to whom specific rules apply.

1.3.3   Interoperability

What had started in 1882 with the Technical Unity was incorporated into European law in 1994 with Directive 96/48/EC of the European Council on interoperability of the trans-European high-speed rail system. This laid the foundations for the creation of a European railway area. Further directives were to follow and concluded in Directive (EU) 2016/797 in 2016, which clearly stated:

"The pursuit of interoperability within the Union rail system should lead to the definition of an optimal level of technical harmonisation and make it possible to facilitate, improve and develop international rail transport services within the Union and with third countries, and contribute to the progressive creation of the internal market in equipment and services for the construction, renewal, upgrading and operation of the Union rail system." [23, Art. 3]

Since 1996, directives have been issued gradually that should lastingly simplify cross-border rail traffic and ultimately create a common European railway area (Fig. 1-3). The term used in this context is the Single European Railway Area (SERA). [24] In 1996 the provision was for high-speed networks only [25], all other (conventional) railway lines followed soon after. [26] In 2008, this distinction was done away with.

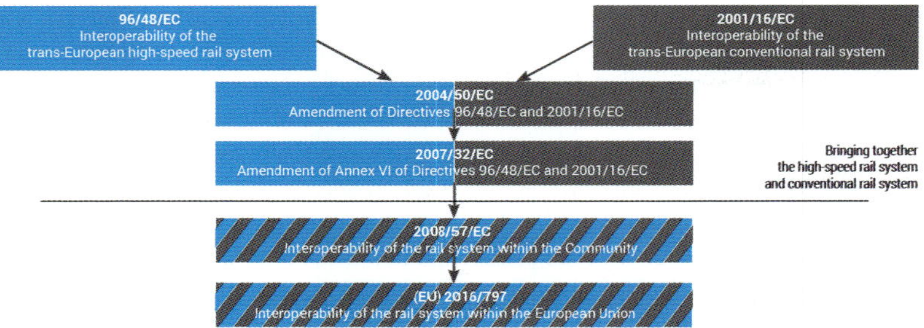

Fig. 1-3: Timeline of key interoperability directives [27, p. 66]

The directives form the legal basis for drafting technical specifications (TSI). TSIs are European Commission regulations. That means that they are valid throughout the EU as soon as they come into force and don't need to be transferred into national law first. The railway system was divided into subsystems in order to push the harmonisation forward. Two of these subsystems (infrastructure and energy) and their associated regulations ([12] and [28]) are of specific interest in this book.

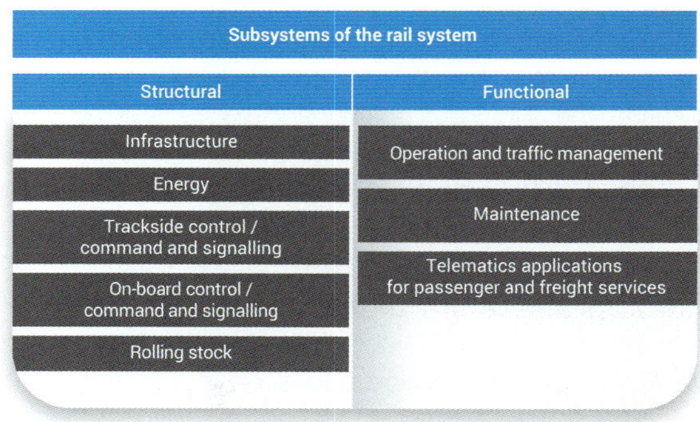

Fig. 1-4: TSI subsystems [27, p. 68]

TSI Infrastructure applies to all types of railway, with the exception of: [29, Art. 1(3)]
– metros, trams and other urban and regional rail systems
– networks that are functionally separate from the rest of the railway system
– privately owned railway infrastructure and vehicles exclusively used by the owner
– infrastructure and vehicles reserved for historical or touristic use

Basic parameters of TSI "Infrastructure"
Line layout
Structure gauge – *EN 15273*
Distance between track centres – *EN 15273*
Maximum gradients
Minimum radius of horizontal curve
Minimum radius of vertical curve
Track parameters
Nominal track gauge
Cant
Cant deficiency – *TSI LOC&PAS, TSI WAG*
Abrupt change of cant deficiency
Equivalent conicity – *EN 15302, EN 13715*
Railhead profile for plain line – *EN 13674*
Rail inclination
Switches and crossings
Design geometry of switches and crossings – *Rolling stock TSI*
Use of swing nose crossings
Maximum unguided length of fixed obtuse crossings
Track resistance to applied loads
Track resistance to vertical loads – *EN 14363*
Longitudinal track resistance
Lateral track resistance – *EN 14363*
Structure resistance to traffic loads
Resistance of new bridges to traffic loads – *EN 1991, EN 1990*
Equivalent vertical loading for new earthworks and earth pressure effects imposed on new structures – *EN 1991*
Resistance of new structures over or adjacent to tracks – *EN 1991*
Resistance of existing bridges and earthworks to traffic loads – *EN 1991, EN 1990*
Immediate action limits on track geometry defects
Immediate action limit for alignment – *EN 13848*
Immediate action limit for longitudinal level – *EN 13848*
Immediate action limit for track twist – *EN 13848*
Immediate action limit of track gauge as an isolated defect
Immediate action limit for cant
Immediate action limits for switches and crossings
Platforms
Usable length of platforms
Platform height – *TSI LOC&PAS*
Platform offset – *EN 15273*
Track layout alongside platforms
Health, safety and environment
Maximum pressure variations in tunnels – *TSI LOC&PAS*
Effect of crosswinds – *TSI LOC&PAS*
Flying ballast ("outstanding item")
Provision for operation
Location markers
Equivalent conicity in service – *TSI LOC&PAS*
Fixed installations for servicing trains
Toilet discharge – *Rolling stock TSI*
External train cleaning facilities
Water restocking – *Rolling stock TSI*
Refuelling – *Rolling stock TSI*
Electrical shore supply – *Rolling stock TSI*
Operating rules
Maintenance rules
Professional qualifications
Health and safety

Fig. 1-5: Key data of TSI Infrastructure [27, p. 70]

In addition to basic technical features of track construction and route layout, the regulation stipulates maintenance rules as well as a definitive maintenance plan as part of overall safety management (see Fig. 1-5). Here, TSI frequently refers to applicable standards in the railway sector which makes the quoted sections legally binding. The expansion of TSIs in the coming years will significantly impact upon or replace the rules and standards of individual railways.

1.4 Cross-border standardisation

Existing legal documents refer repeatedly to further information in standards or quote the state of the art as an equivalent basis for making decisions. In order to harmonise standardisation in Europe, CEN (Comité Européen de Normalisation) was established in Brussels in 1961. While European legislation is in principle only valid for member states, the applicability of the CEN standards goes beyond that. All CEN member states have to adopt the common standards into national standards within a certain transition period. This is the reason why in Austria, for example, up to 90 % of national standards have been replaced with CEN standards. [30] The CEN member states include the EU countries as well as the EFTA countries, Macedonia, Turkey and Serbia. Since 1990, the Technical Committee TC 256 with its title "Railways" has been in existence. In contrast to the legislation described above, the Technical Committee includes all types of railways and therefore was given the title "Railway Applications". [3, p. 17]

There is no universal answer to the question whether standards are legally binding. Sections of standards that are referred to in regulations and laws are considered legally binding. This also applies if individual standards or guidelines and their compliance have been contractually agreed. If there is the requirement to meet the state of the art, standards will assist with this. [4] In substantiated cases or in the case of innovations that are not covered by existing standards it is possible to deviate from existing guide values. This will often result in a necessity for additional verification regarding the safe adoption of the proposed solution.

In addition to the CEN standards, there are also the standards of the International Organisation for Standardisation (ISO), with its headquarters in Geneva. A separate ISO Committee (ISO/ TC 269) named "Railway Applications" was set up in 2012, based on recommendations from France and Germany. [3, p. 31]

In contrast to CEN standards, ISO standards do not have to be adopted as national standards but are accepted as standards in many countries. Although cooperation between the International Organisation for Standardisation and the European Committee for Standardisation was agreed in the Vienna Agreement in 1991, differences are noticeable in some areas. Therefore, it is important to clarify prior to an activity which standards apply in which version.

In the case of the DACH countries, European CEN standards take priority. Apart from standards on the subjects
- manufacturing and joining of rails,
- rail fastening systems,
- sleepers as well as
- switches and crossings,

all standards relating to track geometry and its maintenance are of particular interest. Figure 1-6 shows an extract.

13803	Railway applications - Track - Track alignment design parameters - Track gauges 1435 mm and wider	EN 13803
16704	Railway applications - Track - Safety protection on the track during work - Part 1: Railway risks and common principles for protection of fixed and mobile work sites	EN 16704-1
	Railway applications - Track - Safety protection on the track during work - Part 2-1: Common solutions and technologies - Technical requirements for Track Warning Systems (TWS)	EN 16704-2-1
	Railway applications - Track - Safety protection on the track during work - Part 2-2: Common solutions and technologies - Requirements for barriers	EN 16704-2-2
	Railway applications - Track - Safety protection on the track during work - Part 3: Competences for personnel related to work on or near tracks	EN 16704-3
13848	Railway applications - Track - Track geometry quality - Part 1: Characterisation of track geometry	EN 13848-1
	Railway applications - Track - Track geometry quality - Part 2: Measuring systems - Track recording vehicles	EN 13848-2
	Railway applications - Track - Track geometry quality - Part 3: Measuring systems - Track construction and maintenance machines	EN 13848-3
	Railway applications - Track - Track geometry quality - Part 4: Measuring systems - Manual and lightweight devices	EN 13848-4
	Railway applications - Track - Track geometry quality - Part 5: Geometric quality levels - Plain line, switches and crossings	EN 13848-5
	Railway applications - Track - Track geometry quality - Part 6: Characterisation of track geometry quality	EN 13848-6
13231	Railway applications - Track - Acceptance of works - Part 1: Works on ballasted track - Plain line, switches and crossings	EN 13231-1
	Railway applications - Track - Acceptance of works - Part 3: Acceptance of reprofiling rails in track	EN 13231-3
	Railway applications - Track - Acceptance of works - Part 4: Acceptance of reprofiling rails in switches and crossings	EN 13231-4
14969	Railway applications - Track - Qualification system for railway trackwork contractors	EN 14969
15273	Railway applications - Gauges - Part 1: General - Common rules for infrastructure and rolling stock	EN 15273-1
	Railway applications - Gauges - Part 2: Rolling stock gauge	EN 15273-2
	Railway applications - Gauges - Part 3: Structure gauges	EN 15273-3

Fig. 1-6: Standards relating to track construction and maintenance

1.5 Internal railway guidelines

Despite some differences, every railway act emphasises the importance of safe operation of the railway infrastructure and makes railway infrastructure managers responsible for this. As addenda to the TSI and European standards, the railways with their rules and standards will often have a comprehensive collection of different guidelines. The rules and standards of railways can expand on the details of TSI Infrastructure but in principle must not contradict it. They form an essential contribution to the required safety management and provide much more detail on certain aspects.

This book will mention the guidelines of individual railways and their associated standards in some places. This is to illustrate fundamental relationships and differences but cannot be a substitute for consulting the guideline. Often it will be an extract illustration based on the quoted issue of the guideline or standard. These are continually updated and amended. Therefore, it is always necessary to check the guidelines directly for the consideration of any limit values or before defining operational guidelines.

In Switzerland, the rules of the individual infrastructure operators are compiled in the Regelwerk Technik Eisenbahn RTE (Railway Technical Rules and Standards). While the individual infrastructure managers jointly discuss and agree its contents, the compendium is coordinated and published by the Association of Public Transport (VöV). The RTE documentation "Standard gauge track in practice" is not only a reference book with valuable information and guide values relating to rail infrastructure but also provides cross referencing to the applicable Swiss regulations.

In addition to the rules and standards, there are individual special topics, for example the "track excavation guideline" [31]. Guidelines are published by the BAV and substantiate legal terminology from laws and regulations. In principle, it is possible to deviate from the guideline but this may require special verification.

In Germany, the ELTB is a similar compendium of those rules that are to be applied in principle to the government owned railways. Furthermore, the ELTB details any existing Technical Notes on the guidelines. Technical Notes permit a change to the existing rules and standards without having to adjust the whole guideline. [10] The rules and standards of Deutsche Bahn are divided into various main groups, with chapter 8 "Construction Technology, Control, Signalling and Telecommunications Technology" being of primary importance for tamping work.

In addition to laws, regulations and standards, additional rules and standards apply on the ÖBB network; they are divided into the following groups:

1. Basis for planning
2. Construction technology
3. Vehicle technology
4. Network access regulations for railway undertakings
5. Power technology
6. Control command and signalling, telematics
7. Operation

In order to gain technical approval for the use of track construction machines, the adherence to these rules and standards will have to be proven on a test work site. If work is carried out by subcontractors, the rules and standards from part of the contract and as such are binding.

The UK Government published an implementation plan for compliance with the Infrastructure TSI in 2016 entitled "UK Implementation Plan for Infrastructure TSI". It also publishes a list of specific derogations where a justifiable reason not to comply has been sought. Responsibility for rail industry infrastructure standards lies with the Railway Safety and Standards Board (RSSB). Network Rail, as owner and manager of the mainland rail infrastructure, maintains its own suite of more specific subordinate standards and codes of practice.

1.6 Summary

By way of an introduction to this book we have learned that from an early beginning of railways, with their own rules and regulations, gradually over time an hierarchy of rules and engineering standards came into being which created improved safety and greater opportunities for wider journeys by rail. New European standards, such as the Technical Standards for Interoperability, have broadened these opportunities across a much wider network than might have ever been envisaged when George Stephenson built his Rocket in 1825. In subsequent chapters we shall look at infrastructure rules and standards for track in more detail.

2 Basic principles of the permanent way

Fabian Hansmann and Richard Spoors

2.1 Key issues

The focus of this chapter is the permanent way. It describes its components and their fundamental interdependencies. The emphasis is on conventional track; turnouts are only covered at a rudimentary level. The main track terminology is explained in depth and the following questions are dealt with in detail:

- Which parameters ensure the safe passing of two trains?
- What tasks do the different components of the railway track have?
- How do temperature differences affect the components?
- What are the advantages and disadvantages of different types of sleepers and their materials?
- How is the safe changeover between two tracks ensured?
- What wear phenomena are the components subjected to?
- What different types of turnouts are there and what is their purpose?

2.2 The railway as a system

The railway combines a number of different specialist fields. Different disciplines work together, from classic civil engineering via mechanical engineering to data scientists and IT specialists. An understanding of the systems required for this is of crucial importance. For example, incorrectly implemented lifting values can lead to twist or damage to the overhead line equipment and cause the failure of parts of the system. Individual actions can quickly trigger complete chain reactions at the end of which parts of the system may fail. Although the main focus of this chapter is on the permanent way and its components, the declared objective is to impart an understanding of the interactions in and effects on the system.

Railway infrastructure is generally accepted as being divided into the following main components:

- track
- points
- level crossings
- engineering structures (bridges, tunnels, etc.)
- rail-related elements of stations
- safety and protective equipment

Track, also known as the permanent way, can be described as an engineering system, combining a number of different components which work in the same way across borders. Despite constant harmonisation, different terminology has developed in the countries included in this book. This chapter provides a high level overview of the technical principles behind track design, the manner of its application to the operational railway and an explanation of differing terms.

2.3 Vehicle gauge and structure gauge

Trains running on railway lines cannot avoid obstacles due to the constraints imposed by running on a guiding track. Gauge is about the compatibility between the train and the infrastructure around it. Gauging, therefore, defines an area over the track cross-section which has to be kept clear of all fixed track installations and other fixed obstacles to ensure the safe movement of the vehicle. [1] The absolute gauge results from taking account of allowances for vehicle dynamic movements. It represents the space required to allow for this movement, called the swept envelope. The minimum clearance results from individual measurements to this envelope taking account of the reserves necessary for carrying out track maintenance (maintenance tolerances) [2]. In the UK these track tolerances are based on three stages of track fixity, which is how track position varies with time and use. High fixity comes from slab track; medium fixity from glued ballast and low fixity from normal ballasted track. [3]

As vehicles travel on a route there will be particular places where the clearances have been allowed to be reduced to a minimum. Additional measures will be required at these locations to ensure that clearances are managed. If obstacles, such as trees, project into this space, this can result in a collision with the train so that continued safe operation of the infrastructure cannot be maintained. This provides an additional responsibility upon the infrastructure manager to ensure the safe passage of trains.

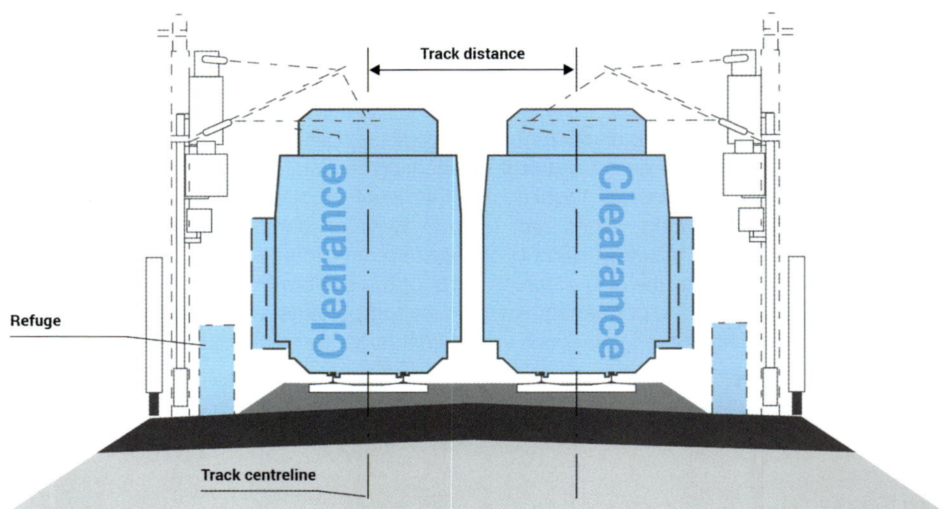

Fig. 2-1: Typical cross-section of a new double track in Austria, based on [4, p. 8]

Based on the considerations of gauge, absolute gauge and minimum clearances, infrastructure managers frequently set the structure gauge specific to a route as standard (standard structure gauge, standard clearance) (see Fig. 2-1). This simplifies planning considerably. Thus, Fig. 2-2 shows the standard gauge in accordance with Appendix 1 of EBO (German Railway Construction and Operations Regulation) Article 9. [5] The different boundaries (small and large boundaries) can be seen as well as areas A and B which are coloured blue. In area A, obstacles are possible permanently if the operation requires this. Classic examples are platforms and signalling equipment. In area B, only temporary installations are permissible during construction work, e.g. safeguarding measures.

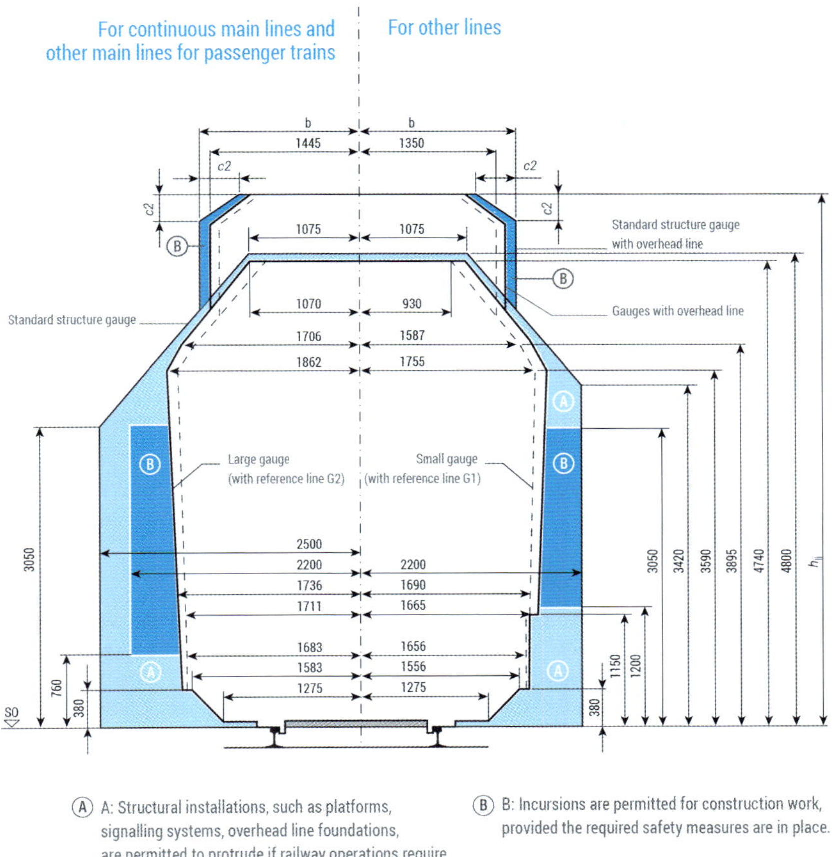

Fig. 2-2: Deutsche Bahn standard clearances in in straights and curves for radii ≥ 250 m [5, Appendix 1 to Art. 9]

EBO Article 9:

"The standard gauge is composed of the space enclosed by the relevant boundary and additional spaces for construction and operational purposes" [5, Art. 9(1)].

EisbBBV (Austrian Railway Construction and Operations Regulation) Article 16:

"The structure gauge is the space which is part of the track and has to be kept clear. The structure gauge is composed of the space required for the unhindered travel of rail vehicles and additional spaces for construction and operational purposes. The structure gauge shall be specified by the railway infrastructure manager on the basis of the relevant kinematic gauge line in accordance with the state of the art; this specification may be dependent on the railway line." [6, Art. 16(1)]

"The minimum structure gauge encloses the space required by a rail vehicle taking account of horizontal and vertical movements as well as track geometry tolerances and minimum distances from overhead line equipment. To calculate the minimum structure gauge, the dimensions of the relevant kinematic reference line shall be extended in accordance with the state of the art." [6, Art. 16(2)]

AB-EBV (Swiss Implementing Provisions of the Railway Regulation):

"The structure gauge [...] is the envelope of the space to be kept clear for the transit of vehicles and other operational purposes. It comprises the boundary of fixed installations and the additional refuges required." [7, AB 18.1(1.1)]

> Refuges shall be provided on both sides of the line and ensure the safety of staff whilst trains are passing. If the track area has to be cleared, e.g. because of a passing train, staff have to immediately move into the refuge.

UK Railway Group Standard GIRT7073 defines structure clearance as:

"The minimum calculated distance between a structure and the swept envelope of a specific type of rail vehicle passing at nominated speeds, taking account of appropriate track tolerances and accuracy of measurement".

On a double track, the rails have to be arranged in such a way that vehicles can pass without danger at the permissible speed.

Similar to the different structure gauges, EN 15273-3 differentiates between the absolute distance between tracks, the minimum distance between tracks and standard distance between tracks. [2]

Furthermore, two terms have to be differentiated:

- The distance between track centres (track distance) in accordance with EN 15273-3 specifies the distance between two neighbouring tracks. [2] This is measured parallel to the running surface of the so-called reference track. The track with the lower cant is regarded as the reference track, with the running surface being an imaginary plane across the top edges of the rails.
- TSI Infrastructure, on the other hand, refers to a horizontally measured distance between track centres in accordance with the applicable legislation in the DACH countries. [8] This is also used in these countries for the implementation of maintenance work. If subsequently reference is made to the distance between tracks, this refers to the horizontally measured distance between the two track axes - unless otherwise specified.

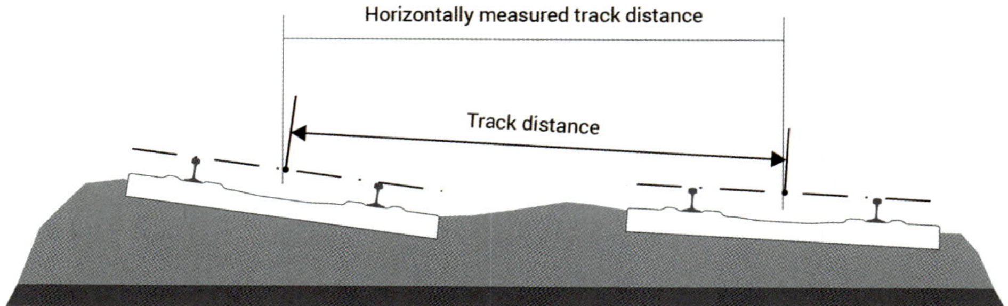

Fig. 2-3: Distance between track centres as per EN 15273-3 [2] and horizontally measured distance between track axes

> The track axis (centreline of track) runs in the longitudinal direction of the track at the level of the top of the rail and has the same distance to both running edges, but without taking account of gauge widening. The centreline of the track coincides with the track axis (reference axis) situated halfway between the two rails.

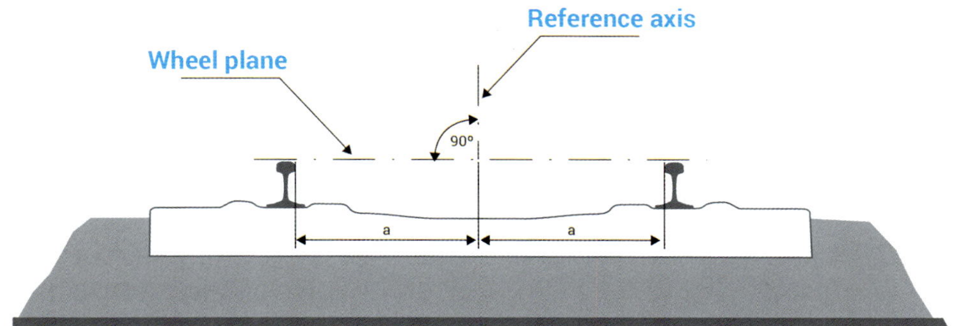

Fig. 2-4: Diagram of the track axis and the running surface as per EN 15273-1 [9]

National legislation provides the following values for the standard horizontal distance between track centres (track distance):

	Germany	Austria	Switzerland	UK
	EBO Art. 10	EisbBBV Art. 17	AB-EBV, AB 19.1 ff	Network Rail
Plain line	4.0 m (3.8 m*)	4.0 m	3.8 m	3.405 m
Station area	4.5 m	4.5 m	4.5 m (4.2 m*)	3.405 m
Addendum	Adjustment for radii smaller than 250 m	Adjustment for radii smaller than 250 m	Adjustment for radii smaller than 250 m	N. Rail does not measure track centres but the spacing between tracks at 1.970
Railway directive	RIL 800.0130	RW 01.03	RTE 20012	NR/L3/TRK/2049/MOD07

* Exemption

Fig. 2-5: Generally applicable distances between track centres on plain track for new lines

In Austria, a track distance of 4.00 m is required in the ÖBB network on plain track for speeds below 160 km/h. This distance is increased to 4.50 m for higher speeds. [4]

In the DB network, a distance between track centres of 4.50 m [10] applies to new and upgraded lines with a design speed of more than 200 km/h. In principle, the distance required is announced before maintenance work is carried out. To simplify the measurement in practice, the distance from the left-hand rail of the adjacent track to the left-hand rail of the reference track may be stated (inside edge to inside edge).

Gauge in the United Kingdom

The main line railways of the United Kingdom have historically had a smaller structure gauge to those of mainland Europe. There was no mandated standard until 1846 when the Railway Regulation (Gauge) Act was introduced. However, it only defined the track gauge, and left individual railway companies to define clearances. In general these were as small as practical to minimise the cost of construction of new lines, leaving the UK with one of the smallest structure gauges in the world.

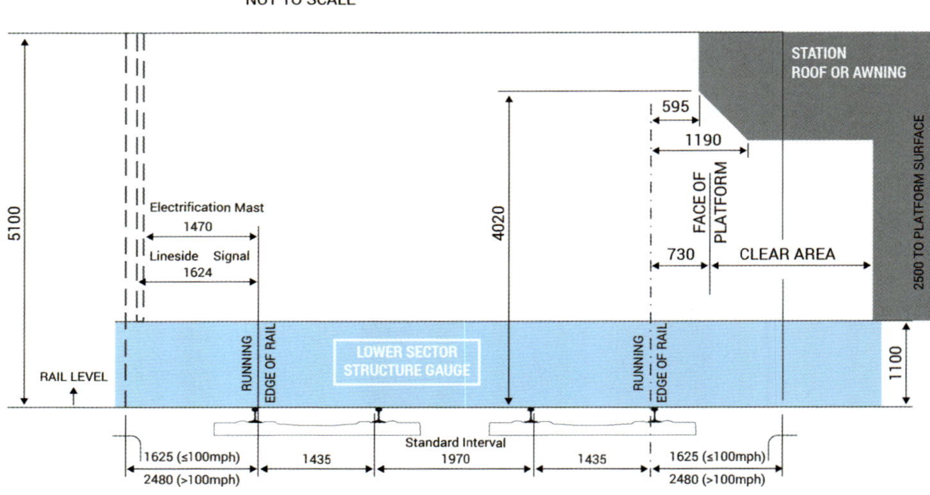

Fig. 2-6: UK standard structure gauge showing principal dimensions for new lines. Note the shaded area that shows the lower sector. [11]

The description of the management of vehicle gauge and structure gauge in the DACH countries, together with EN 15273, is based on the principle of reference profiles. They represent a notional boundary between the space occupied by rolling stock and the space that may be occupied by infrastructure. In the United Kingdom, railways have grown in popularity and usage over the last 25 years. During that period there has been a significant increase in new trains to meet demand and expectation, including tilting passenger trains and larger freight vehicles carrying containers. This has produced a challenge to the management of the infrastructure as seemingly larger and longer trains want to operate on tracks that have always been close to platform edges and bridge soffits. The industry, therefore, has developed the management of gauging into an extremely sophisticated series of processes using state of the art measuring equipment and computerised analysis tools. Gauging in the UK can be described as the systematic understanding of the component systems, followed by the application of safety and management constraints, to ensure that the stated clearance between the track and the physical infrastructure is maintained. The safe passage of trains is managed on a route by route basis, which means that for certain vehicle types they are restricted to particular routes. This is especially the case with freight vehicles, where certain routes have been especially 'cleared' for their journeys. In the past, when gauging was based on reference profiles, the physical clearance was adequate to allow for a degree of encroachment by the vehicle body or the maintenance of the track position into that space. What has changed is the definition of clearance for the infrastructure manager. These are 'normal', (greater than 100 mm) 'reduced' (greater than 50 mm) and 'special reduced' (greater than 0 mm). Track has also been defined with regards to its lateral fixity, with each definition having a dimensional tolerance which must be taken into account when measuring and calculating the clearance. These are high (slab track), medium (glued ballast) and low (ballasted track) with respective tolerances of 0, +/-15 mm and +/-25 mm. A vertical tolerance of up to 25 mm is allowed in both medium and low definitions. As the whole process has become more precise, the controlling Railway Group Standards have separated gauging vertically to the lower section of the infrastructure, below 1100 mm and the upper section. [3]

The opening of the Channel Tunnel in 1994 and the building of HS1 from Folkestone to London St. Pancras International brought the UIC GC gauge into the UK and HS2 from London to Birmingham and the north is expected to be built to this gauge.

2.4   Axle loads on the track and their distribution

Over the years, the loading of the track has changed significantly. Increasing axle loads, speeds and extreme weather challenge the track every day anew. Longitudinal, lateral and vertical forces arise in the track due to stresses from weather and traffic, and the track system must counter these effects.

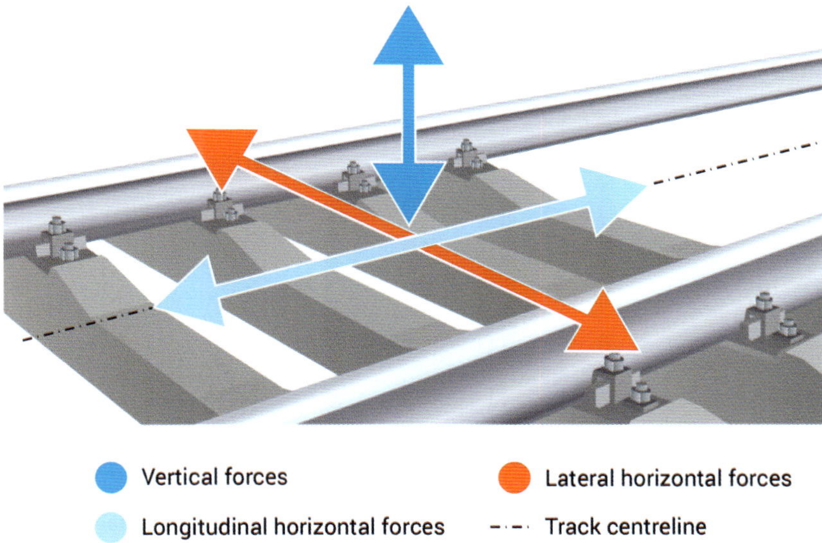

Fig. 2-7: Main directions of forces in the track

The system must absorb and distribute the applied forces. Depending on the components used, it is possible to absorb different loads, and the type of track construction must be matched accordingly. The system counters the stress with different resistances which permit absorption and transmission of the forces. If the forces exceed a specific magnitude or act on the system for a prolonged period, individual components start to wear or fail as a result of fatigue. The result of this may not be just a failure of parts, but the complete track system.

Various maintenance measures have been developed to maintain the functional condition of individual components. It is the task of the infrastructure manager to monitor the wear process and through timely adoption of these measures enable a long-term and trouble-free use of the asset.

2.4.1 Vertical forces and deflection

Vertical forces arise from railway traffic and, only in exceptional cases, e.g. at railway crossings, by loading from other road users. They have a static and a dynamic component and are introduced into the system via the contact face between wheel and rail.

In the DACH countries, the permissible axle loads on mixed traffic lines are typically 22.5 t (depending on the relevant line category). Including safety margins, the above figure results in the static force to be assumed, i.e. that force which acts on the track when the train sits on it without moving. In the UK the maximum permissible axle load is 25.5 t.

Mixed traffic lines are railway lines on which both freight and passenger services run. In contrast to pure high-speed lines, these lines, on average, are used with higher axle loads.

When the vehicle moves, the acting force increases with increasing speed due to the dynamics of the vehicle. The speed of the vehicle, its running behaviour, unsprung masses etc. affect the energy introduced. This phenomenon, as a basic principle of dynamics (kinetic energy), can be compared to driving a nail into a board. Driving a nail into a board will hardly be successful if one simply places the hammer onto the nail. However, due to the swinging motion, it is possible to transfer much more energy onto the nail. In the same way, the static axle load is increased by the vehicle's movements.

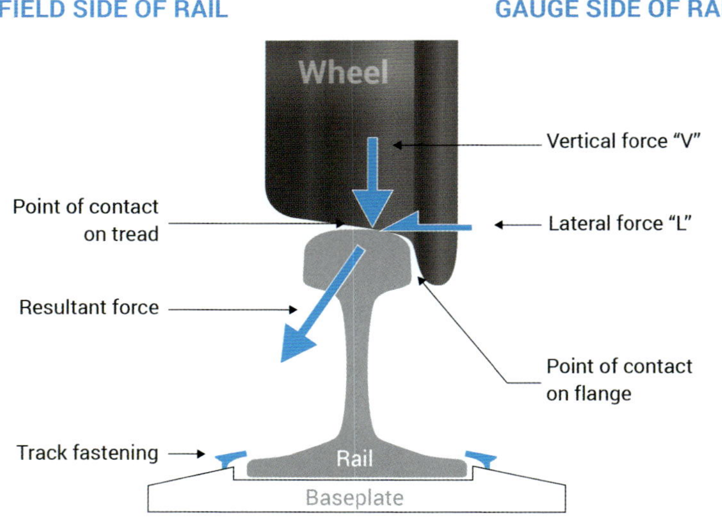

Fig. 2-8: Forces acting at the wheel/rail contact patch

Vehicle and track interact. The structure of track has a major influence on the way the load is transferred. At rail joints and other irregularities at the wheel-to-rail contact point (e.g. in case of rail faults), a pulse-type excitation of the vehicle takes place. This greatly increases the stress on the total system. Sleepers which are asymmetrically supported or tamped unevenly have a similar effect.

On the road, this phenomenon can be experienced by a vehicle when hitting a pothole. If there is a small amount of damage to the road surface after winter, it will quickly increase. A small unevenness increases the effect on road and vehicle. The road, which is already weakened, is stressed more severely and thus wears more quickly.

Although the loadings on the track of high-speed, heavy freight and mixed traffic are different, they can be compared with regard to their fundamental effect. The system must not only withstand them but ensure sufficient availability in addition to safety. The contact area between wheel and rail is hardly bigger than a fingernail. Considering that loads of several tonnes are transferred across such a small area, one can imagine the large stresses arising from this. It is, therefore, necessary to absorb the forces as well as possible and transfer them through the individual components.

The track system has an inherent elasticity which helps in the load transfer mechanism. The elastic deformation of track under the action of the wheel load permits the distribution of the load across several sleepers. The track acts as a sort of mass/spring system which transfers the load from a fingernail-sized area to a much larger one via this deflection. A common method of reducing the acting force is by increasing the load bearing surface area. For example, snowshoes make it possible in this way to travel over snow without sinking in. Without snowshoes, the bodyweight would act on a much smaller area and one would sink into the snow. Rails work in a similar way, by deflection, which is the fundamental principle of load transfer. The deflection, which can vary between 0.7 and 2 mm for ballasted track, depends upon the materials used, the substructure and the load applied. In this way, it is possible to transfer the load across several sleepers (5 to 9 sleepers) and into the ballast. A single sleeper would only be capable of counteracting such an enormous load for a limited time. There are different mechanisms for non-ballasted track, which are not considered in this book.

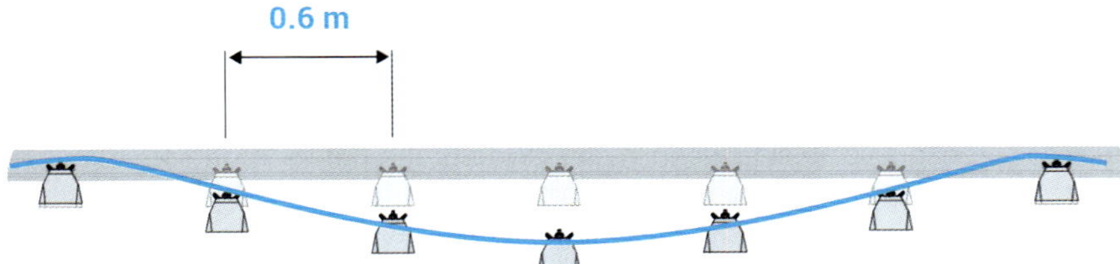

Fig. 2-9: Load transfer by means of rail deflection

The load is transferred to the ballast bed via the contact area between sleeper and ballast. A large part of the load is passed directly into the ballast bed via the supporting surface created by tamping directly underneath the rails. The load is transferred through the ballast with a specific load-spreading angle into the lower layers. The loads are transferred to increasingly larger areas and thus gradually reduced, as the diagram below shows.

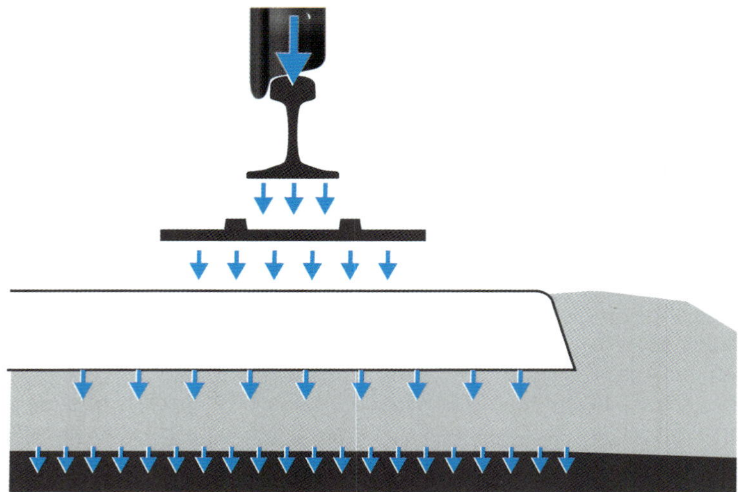

Fig. 2-10: Load transfer and how it spreads from the wheel down into the formation

2.4.2 Lateral forces

Lateral forces arise in the track for different reasons:
- steering
- vehicle traction and braking
- vehicle oscillations
- wind effect

The steering effect of vehicle wheelsets generates horizontal force components on both straight sections and in curves. These forces are markedly larger in curves and can lead to considerable additional stresses on the components. Depending on the cant of the track and the speed, the vehicle is either guided along the outer rail of the curve by the wheel flange, moves over the rail in equilibrium or slides towards the inside of the curve onto the inner rail.

The lateral forces can lead to side wear on the inside of the outer rail, the rail having a tilting movement (twisting) and a displacement towards the end of the sleeper (expansion force), independently of their origin. The rail fasteners act against both forces. A widening of the track, which would otherwise be possible, could lead to the derailment of the train and would have devastating consequences. The lateral forces are absorbed like the vertical forces and transferred to the ballast via the sleeper where the lateral resistance of the ballast bed takes effect. This consists of the following components:

- **Sleeper base resistance**
  The classic friction resistance builds up between the ballast and bottom of the sleeper against a possible displacement. This resistance has different magnitudes depending on the engagement with the ballast bed, the sleeper type and the weight. The resistance of the sleeper base only acts if the track is loaded with at least its own weight. If the track is lifted even fractionally (take-off wave during movement of train), the resistance is reduced significantly.

- **Sleeper side resistance**

  Sleeper side resistance is defined as the resistance between the ballast in the sleeper crib and the side of the sleeper. It can vary depending on the shape of the sleeper, the type of ballast and its consolidation.

- **Ballast shoulder resistance**

  When the sleeper starts to move outwards, the weight of the ballast at the shoulder counteracts this movement. The area of ballast between the sleeper end and the end of the top of the ballast (start of the slope) is called the ballast shoulder.

Furthermore, different types of sleeper significantly affect the size of the lateral resistance due to their weight, the shape of the sleeper and the size of the contact area (more on this in section 2.6). As for vertical deflection, a system of several sleepers acts together in the case of the lateral forces (framework effect). This interaction greatly increases the resistance to applied loads and creates necessary safety margins. Nonetheless, the effects of temperature stresses, in particular, should not be underestimated.

2.4.3   Longitudinal forces

Steel expands when subjected to heat, which is of particular interest for rails as linear structures. During a year, the rail is subjected to extreme temperature fluctuations. In summer, in particular, the rail temperature can increase to significantly above 50 °C when directly exposed to the sun. The resulting change in length of a rail supported frictionless is as follows:

$$\Delta l = l \cdot \alpha \cdot \Delta t \tag{2-1}$$

Δl in [m]:   change in length of rail

l in [m]:   rail length

α in [m/m]: coefficient of thermal expansion of rail steel 0.0000115

Δt: temperature difference to temperature when laid

If a change in the length of the rail is not possible, longitudinal forces arise internally. In winter, this can result in an increase in rail fractures and in summer to distortion or buckling of the rail.

A brief calculation:

For a 45 m rail supported frictionless, a change in length results as follows:

$$\Delta l = l \cdot \alpha \cdot \Delta t$$

$$l = 45 \text{ m} \qquad \Delta t = 20\,°C$$

$$\Delta l = 45 \cdot 0.0000115 \cdot 20$$

$$\Delta l = 0.01035 \text{ m}$$

The rail expands by about 10 mm due to an increase in temperature of 20 °C with frictionless support. The temperature increase in this case does not refer to the ambient temperature but to the temperature of the rail. This can be considerably higher than the air temperature due to solar radiation. Nowadays, rails are rarely connected by means of separable rail joints. They are welded together and thus form longer rail sections. Moreover, the rail is not supported on the sleeper in a frictionless manner. Thus, the change in temperature does not cause a change in length but leads to stresses forming in the rail. These can be calculated as follows:

$$\sigma = \alpha \cdot E \cdot \Delta t \tag{2-2}$$

σ in $\left[\dfrac{N}{mm^2}\right]$: Stress in the rail

E in $\left[\dfrac{N}{mm^2}\right]$: Young's modulus for steel $2.1 \cdot 10^5$

$$\sigma = 0.0000115 \cdot 2.1 \cdot 10^5 \cdot 20 = 48.3 \; \frac{N}{mm^2}$$

Taking into account the cross-sectional area of a 60E1 rail profile (76.70 cm²), this results in a force of 370 kN. This corresponds to about 37 t acting upon it. The rail is exposed to pressure. If the temperature increases further and overcomes the resulting force, i.e. the counteracting resistances, the rail will yield to the force and buckle.

Rails are supplied to the construction site in defined standard lengths, and there they are connected into a continuous rail. This can be done in two different ways. The rails are either joined by fishplates or are welded together. In Europe, it is now state of the art to weld rails and thus produce a continuous welded track. Fishplates may continue to be used, but only in tight curves.

For rails with a joint gap, the temperature stress leads to a change in length and thus to the closure of the joint gap after the counteracting resistances have been overcome. Only then do constraining forces arise in the rail, which can lead to distortion. In continuous welded rails, all temperature changes are immediately converted into stresses. A temperature increase thus leads to compressive stresses as the rail is prevented from changing its length. In contrast, low temperatures in winter result in tensile stress due to contraction of the rail, which can cause rail fractures. In curves, an additional lateral force develops from the longitudinal forces arising due to the curvature of the rail.

Track distortion means a shifting of the rails due to large internal longitudinal compressive stresses which are caused by forces acting from the outside (e.g. take-off wave in front of a train). Track distortion can occur over lengths up to 20 m and can shift the track panel up to one metre in the horizontal direction.

Track buckling refers to a shifting of the track due to large internal longitudinal compressive stresses which are exclusively caused by high temperatures without an external force. Track buckling is indicated by bulging of the track panel and is usually much smaller compared to track distortion, but still safety critical.

In some countries there is less distinction between a buckle and distortion and just the term track buckle is used.

There are various resistances acting in the track panel which counteract a change in length of the rail. These can be divided into the following:

- **Fishplate frictional resistance**
  The fishplate frictional resistance only occurs in jointed rails and is mainly affected by the tightening torque of the fishplate bolts and the condition of the fishplate joint.
- **Resistance to longitudinal displacement**
  If the rail joint is closed, the rail starts to move the sleepers lengthwise. The resistance the ballast offers to this displacement is called the resistance to longitudinal displacement and largely depends on the ballast properties, the ballast profile and the degree of consolidation.
- **Creep resistance or torsional resistance**
  The rail is fixed to the sleeper based on the fastening system used. The resistance varies depending on the size of the holding down force of the fastening system and the friction between sleeper and rail. The use of additional rail-anchoring devices (see section 2.6.3) and a reduction of the sleeper spacing can increase this resistance. The rail can move relative to the sleepers only when the creep resistance has been overcome.

If the rail temperature exceeds a critical value, activities which reduce the resistance of the ballast must be avoided.

Germany [12]

Tamping and lining, loosening of fastenings and clearing of ballast on a track section may only be carried out at temperatures between +3 °C and +35 °C in Germany for continuous welded track. Exempted from this is track of the former German Reichsbahn if it was formed by closure welding between +12 °C and +17 °C. In these cases, such work is not permitted at temperatures above +30 °C due to the altered state of stresses. For track with a joint gap the limit is +25 °C. If the rail temperature exceeds +45 °C, all work has to be discontinued on the following day, even if the rail temperature falls below +35 °C.

Austria [13]

In Austria, tamping of track and switches, loosening of fastenings and clearing of ballast must be stopped from a rail temperature of +38 °C and has to be postponed to night-time or early morning. To prevent track distortion, the development of the weather during the next few days is taken into account to be able to take preventative measures, depending on track and line class, even for temperatures below the limits specified.

Switzerland [14]

Maintenance work requiring removal of ballast or lifting of the track is not permissible for tensioned track at 5 °C above the neutralisation temperature. At temperatures below +5 °C, curves with radii of less than 800 m must be preheated before tamping is carried out. Track at temperatures between +20 °C and +30 °C which has not been neutralised must be separately monitored following the maintenance work.

United Kingdom [15]

In the UK, Network Rail stipulates that maintenance tamping must not be carried out when the rail temperature is at or exceeds +32 °C, nor when it is at or likely to fall below -7 °C. All CWR track is documented in a register that records its stress free temperature (SFT). The SFT shall be in the range 21 °C – 27 °C.

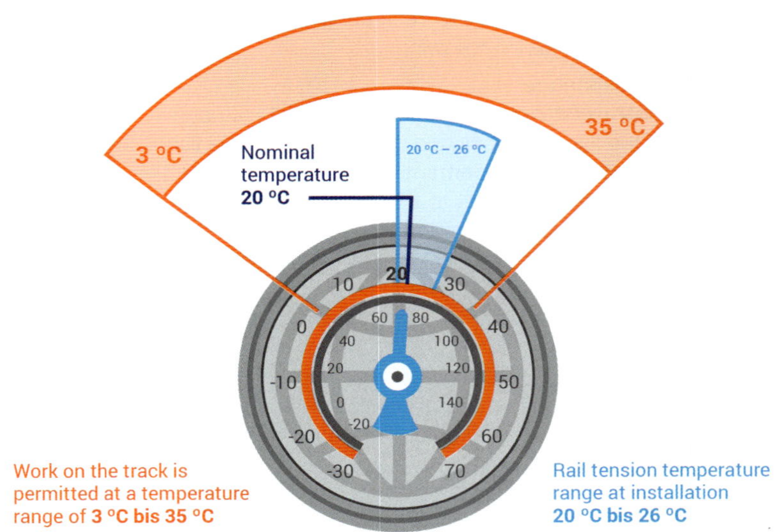

Fig. 2-11: Diagram of the limits for tamping and lining work in the conventional rail tensioning range based on the example of Germany

2.5 Classification of lines

Infrastructure managers define track in different categories to be able to react to different effects in an optimum manner. These categories are differentiated not only by the combination of the components used in the track and substructure but also, amongst other things, by the design of the cross-section.

In general, the classification will be determined by the local loading (load tonnes per day (Lt/d)) and the running speed (km/h) on the section. The track material used, the cross-sectional parameters and the standard values vary depending on these constraints. In special areas, additional parameters, such as the curve radius, may also be taken into consideration when selecting components.

In Germany, unlike in the other DACH countries, the different track categories are directly mapped to the input parameters of track loading ($\leq$ 10,000 Lt/d, > 10,000 Lt/d to < 30,000 Lt/d, $\geq$ 30,000 Lt/d) and speed ($\leq$ 80 km/h, 80 km/h < v $\leq$ 120 km/h, 120 km/h < v < 160 km/h, v = 160 km/h and v > 160 km/h) (see Appendices A04–A06 of RIL 820.2010 [16] – [18]). This mapping will be discussed further in the following sections.

The DB network differentiates between continuous main lines, main lines and secondary lines. The following applies in accordance with EBO Article 4(11):

"Main lines are lines which are utilised by scheduled trains. Continuous main lines are main lines on an open track and their continuation into stations. All other lines are secondary lines." [5]

On the ÖBB network, the line sections are classified according to running speed, track load and type of traffic (passenger traffic and goods traffic). However, a track category based on the classification is not mapped directly to the line sections. Unlike in Germany, each line section is given a track class and a line class in advance. The different rules and standards then use the clearly defined classification to be able to make statements specific to a situation.

The track classes are as follows:

- Track class a: line track and continuous main lines
- Track class b: other main lines
- Track class c: secondary lines

The following applies in accordance with EisBBV Article 11(8):

"Main line tracks are tracks which are equipped with safety systems for train running. Continuous main lines are main lines on plain track and their continuation into stations. All other lines are secondary lines." [6]

Together with the line classes shown in Fig. 2-12, they form an unambiguous combination for the complete set of rules and standards.

Class	Average daily track loading [tonnes/day]	Type of traffic	Notes
S⁺	–	Passenger and freight	all lines (sections) at $v \geq 200$ km/h
S	> 30000 [1] or [2]	Passenger and freight	[1] and $v < 200$ km/h [2] Lines with activated vehicle body tilt
1	> 10000 [3]	Passenger and freight	[3] If they are not class S
2	3000 – 10000	Passenger and freight	
	3000 – 10000	Passenger only	
	> 10000	Freight only	
3	< 3000	Passenger and freight	
	< 3000	Passenger only	
	< 10000	Freight only	> 4 trains/day
3G	–	Only freight and/or nostalgic trains	approx. 4 trains/day

Fig. 2-12: Definition of line classes in the ÖBB network as per [19]

In Switzerland, the division is by track loading groups taking account of track load (total gross register tons per day (t.g.r.t/d)) and target line speed (km/h) among other things. The abbreviations in front of the number are N for new build and E for maintenance projects.

In Switzerland, station track which is entered and departed via signals is regarded as main line track; all others are regarded as secondary lines. [20, section 3.2]

Designation	Future loading	Or future target line speed (train series B)
N1, E1	> 30,000 t.g.r.t/d	or $v \geq 160$ km/h
N2, E2	15,000 – 30,000 t.g.r.t/d	or $v \geq 80$ km/h
N3, E3	5,000 – 15,000 t.g.r.t/d	–
N4, E4	< 5,000 t.g.r.t/d	–

Fig. 2-13: Classification of Swiss track loading groups [7, AB 25(2.2)]

In the United Kingdom, Network Rail assesses each line of route on 3 key factors:

- The speed required on the line, which is the speed required to suit current and future known traffic flows.
- The annual tonnage moved over it, which is the type of traffic that moves over a line and takes into account both passenger and freight trains. Heavy freight trains can cause more damage to the track compared to say a modern day lightweight passenger rolling stock.
- The equivalent tonnage on it, which is Equivalent Million Gross Tonnes per Annum (known as EMGTPA). It is a measure of the annual tonnage carried over a section of track but takes into account variations in track damage caused by different types of rolling stock.

Independent of route, track is assigned a category from 1A to 6 based on a function related to EMGTPA. Category 1A is the highest – 125 mph or higher and Category 6 is the lowest – 20 mph and below. [21]

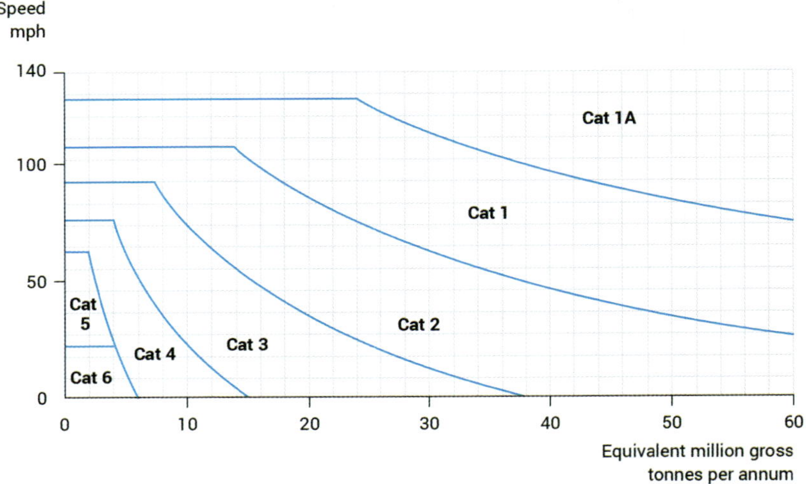

Fig. 2-14: The 7 classifications of Network Rail's track [22]

2.6 The permanent way and its constituent parts

The emphasis of this book is on the permanent way and its extended sphere of influence. The permanent way is composed of different components which take up, distribute and remove various forces from trains using it as an interacting system as described in section 2.4. A successful management of this complex system is necessary to ensure that the demands made upon it with regard to safety, availability, load capacity and suitability for use are met.

A cross-section through the permanent way clearly shows the main constituent parts, the characteristics of which are described in more detail below.

The permanent way consists of the track panel, ballast, substructure and subsoil. The separation between track and substructure (if not present also subsoil) is referred to as "Planum" (formation) in Austria and Germany. In Switzerland, a distinction is made between the transitions. The transition between track and substructure is called "Planie", and the boundary between substructure and subsoil is called "Planum".

Fig. 2-15: Overview of the constituent elements of the permanent way

The track panel and the ballast bed underneath are part of the track system. The track panel consists of the rails, the fastening system and the sleepers, with no distinction being made between plain line and turnouts. Additional layers of sand, gravel or asphalt can be used underneath the ballast bed to improve the load bearing properties of the permanent way. In Austria and Germany, these layers are regarded as part of the track bed, while in Switzerland and the UK they are part of the substructure. Engineering structures such as bridges and tunnels as well as cuttings and drainage systems are also part of the substructure.

2.6.1   The rail

There is a lot of argument about the origins of the railways. It is a fact that mine cars were moved on a type of wooden rail in German and English mines as early as the 16th century. [23, p. 8]

In the 17th century, the idea arose in Nottingham, England to use these wooden tracks outside of the mines. In 1603, a coal mine owner by the name of Beaumont joined the two "wooden rails" with crossbars and thus laid the foundations for the modern railway track. This idea made it possible to increase the volume of materials carried up to 5-fold. The wooden rails were fitted with iron fittings at particularly heavily stressed points to increase the life of the track. [24, p. 1]

The contact of steel on steel between the wheel and the rail underneath is the crucial advantage of the railway system. Due to this low frictional value, less energy is required to transport large masses from A to B.

The original iron fittings on wooden rails were, however, not able to withstand the increasing loads. In 1767, the English ironworks owner Richard Reynolds was, after a few unsuccessful attempts, the first to introduce the flat rail made of cast iron (see Fig. 2-16). [25, p. 7] Cast iron, however, is particularly brittle and not ductile like modern rail steel. Therefore, his flat rails were continuously supported by means of longitudinal timbers. The development of railways in England owed much to George Stephenson, a colliery worker from Northumberland, who studied and learned to repair stationary steam engines as a young man. By 1824 he had opened a factory to build steam locomotives for new colliery railways. This expansion of the early railways also saw the development of foundries and techniques to roll early wrought iron rails. The low level of friction in the contact of metal wheel on metal rail is the crucial advantage of the railway system, as less energy is required to transport large masses from A to B. [26]

Fig. 2-16: Reynolds rail as a representative of the first flat rails, based on [27, p. 16]

Fig. 2-17: Use of L shaped cast iron rails and stone blocks on the Denby Tramway-wagonway (Outram's Railroad) near Little Eaton, Derbyshire, England [28]

The development of new iron rolling mill methods in the UK created stronger wrought iron rails and sleepers reverted to lying crossways. The American R. L. Stevens sailed to England in 1830, bringing with him his ideas for a flat bottom rail section which he had produced in the Dowlais Ironworks in Wales and shipped back to the USA. At the same time Charles Vignole, an Irish engineer, also designed a flat bottomed rail which laid the foundation for the development of today's common rail profile, which was first used in its current shape in 1890. In England the preferred profile was not flat bottom but bull head, a design that was developed from the double headed wrought iron rails and it prevailed until 1948.

Fig. 2-18: Original malleable iron rails and stone block sleepers from the Stockton and Darlington Railway 1825 [29]

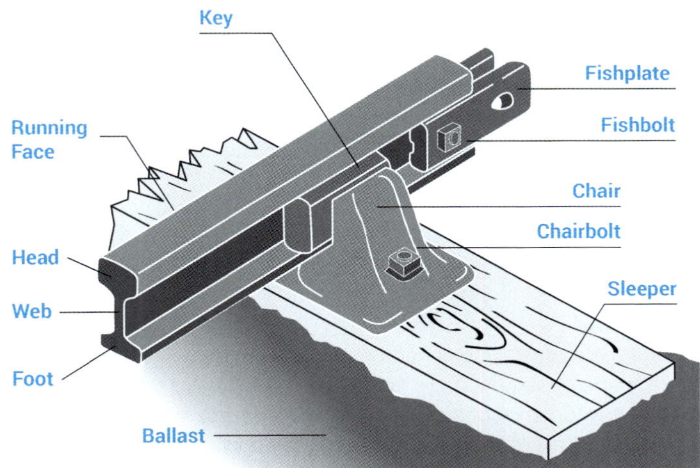

Fig. 2-19: Bull head rails were the standard rail in the UK until 1948 (View is from the field side [30])

In addition to the classic Vignoles rail (see Fig. 2-20), other forms such as the grooved rail for trams were developed outside the standard gauge railway. Vignoles rails basically consist of a horizontal foot, a vertical web and a special rail head. The special shape of the rail and the different dimensions enable the perfect adjustment to the required tasks of load capacity and cost minimisation. It remains a quirk of history that Vignole's rail is more well-known than that of Stevens.

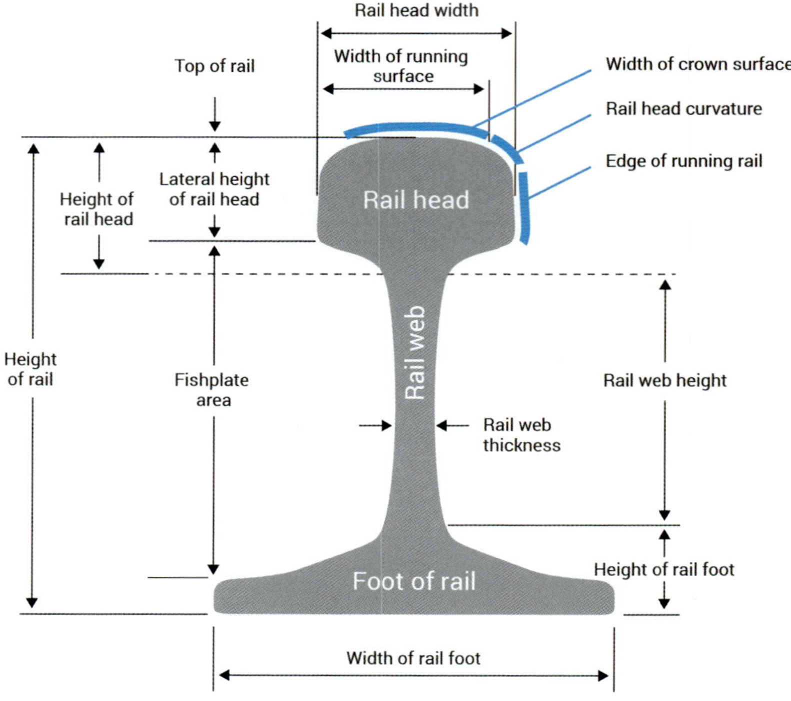

Fig. 2-20: Modern flat bottom rail cross-section based on [31, p. 19]

The rail has three basic tasks:

- **The rail as a pathway**
  The interaction of the matched wheel and rail enables railway vehicles not only to transport masses with little energy expenditure, but also enables easy running through curves due to their geometrically optimised profiles. It is not just the wheel flange which holds the wheelset on the track in tight curves, but also the coned wheel profile which forces the wheelset to centre itself. It is this so-called 'oscillatory motion' or hunting which enables rail vehicles to enter and pass through curved tracks.
- **The rail as an engineering component**
  The rail takes on important mechanical engineering functions. It needs to have a head depth to withstand wear from train wheels; a clever geometric head profile to help keep the wheel contact patch at its centre; a cross sectional shape to withstand and transmit bending and shear forces into the track system and a have a foot section broad enough to both withstand roll-over and spread the load over the support area.

The special shape of a rail defines its geometrical moment of inertia and this has a significant effect on the stresses formed in the rail and the deformation characteristics of the track system.

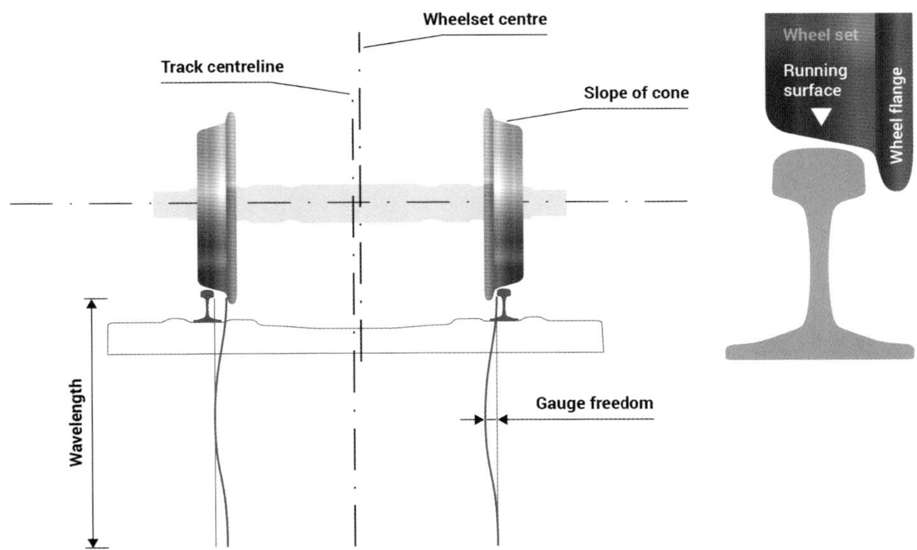

Fig. 2-21: Interaction between wheel and rail as the cause of hunting, based on [32, p. 4]

– The rail as an electric conductor

For railways that have electric traction, the rail is used as a return-current conductor. Furthermore, it can also be used in the control and signalling of trains through the use of track circuits.

The special shape of the rail is due to the particular requirements imposed upon it, and even if the general design is similar, rail profiles differ in detail. In Europe, the uniform designation for Vignoles rails from 46 kg/m upwards is made up of the weight per linear metre and the specific designation of the rail profile in accordance with EN 13674-1. [33] Different rail profiles with the same weight per linear metre are uniquely marked with the letter E and continuous numbering. The height of the rail web and the width of the rail foot, in particular, can vary significantly. In the four countries, only the following rail profiles are currently used for new build and maintenance interventions:

DB	ÖBB	SBB	NR
60E2	60E1	60E1	60E2
54E4	54E2	54E2	56E1
49E5	49E1	54E1	49E1
		46E1	

Fig. 2-22: Standard rail profiles for new build and maintenance interventions [14], [16] – [19], [34]

Profile	kg/m	Height [mm]	Width of rail head [mm]	Width of rail foot [mm]	Rail web thickness [mm]	Moment of inertia x-x (cm⁴)	Modulus of resistance x-x (cm⁴)
46E1	46.17	145	65	125	14	1641.1	217
49E5	49.13	149	67	125	14	1799.7	237
49E1	49.39	149	67	125	14	1816	240.3
54E2	53.82	161	67	125	16	2307	276.4
54E1	54.77	159	70	140	16	2337.9	278.7
56E1	56.3	158.75	69.85	140	20	2321	421.6
60E2	60.03	172	72	150	16.5	3021.5	330.8
60E1	60.21	172	72	150	16.5	3038.3	333.6

Fig. 2-23: Geometric parameters of selected standard profiles as per EN 13674-1 [33] [35]

Many different rail profiles are used in the existing networks and also in shunting yards. In the UK that includes bull head rails. However, due to the wide range of profiles, they cannot be listed individually here. It is imperative to note that the height of different rail profiles can have a significant effect on the overall height of the track. In Europe, rails with a weight of 60 kg per linear metre are standard. The rails of heavy-duty railways in coal and ore mining areas can easily reach a weight of 70 kg per linear metre. Line sections with different rail profiles are connected either by transition sections (profile transition rails) or thermite transition welds.

Besides the rail profile, the rail material is a major factor for success. In addition to high resistance to the loads acting upon it, the steel must meet further criteria. Amongst others, these include resistance to brittle fracture at low temperatures and the capability of being welded to produce a continuous running surface. Over time, these various requirements have resulted in equally numerous developments.

Iron and slag are produced from iron ore, lime and coke in a blast furnace under a continuous supply of fresh air. By means of desulphurisation and decarburisation with pure oxygen, iron is converted to steel. The carbon content of the steel is lowered to below 1.5 %, and the iron is prepared for the subsequent rolling process by possibly adding other substances.

Iron forms the main component of rail steel with 97 to 98 %. The remainder includes other chemical substances, such as carbon, manganese or silicon. Despite their small percentage, individual additives or impurities can affect the properties of rail steel. [36, p. 67]

The secret of rail steel is its structure. The carbon component plays an important role. If its percentage in the rail increases, hardness and strength increase. However, the resistance to brittle fracture is reduced and the possibility of sudden material failure due to overload increases, unless the ductility required is produced by additional processes (e.g. heat treatment). Due to its low carbon content, rail steel can absorb large forces through plastic deformation before breaking.

In contrast to elastic deformation, permanent deformation of the steel occurs under high loads. This remains even after the load has been reduced and is described as plastic deformation. Materials with low ductility cannot deform plastically under load and break immediately. A good example of this is glass which, unlike steel, cannot deform under load and breaks quickly.

The forces arising in wheel/rail contact are massive. Thus, there are continuous attempts to increase the strength (hardness) of the rail and its resistance to these forces, without reducing its positive properties, such as weldability and ductility. In Europe, the standard rail quality is generally R260 with a Brinell hardness of 260 – 300 BHN, although high-strength and super high-strength steels (e.g. R350HT and R400HT) are increasingly used.

The hardness test for rails is carried out in accordance with EN 13674-1 based on the test method developed by the Swedish engineer Johan August Brinell and named after him. [33] During this materials test, a carbide ball is pressed vertically onto the test specimen with a defined test force. The diameter of the indentation in relation to the diameter of the carbide ball gives the Brinell hardness (BHN).

Heat treatment or the addition of alloys, such as chromium, molybdenum, vanadium and titanium [36, p. 67], make it possible to improve the steel hardness and other properties of the material without significantly affecting the positive basic properties. The wear of the rail profiles in curves, in particular, can arise from high wheel forces. This wear can affect the gauge, the running characteristics and the load capacity of the rail. With increasing loads in rail traffic, further issues have arisen.

These are associated with the development of rolling contact fatigue cracking (RCF) in the rail head that, if left untreated, can lead to rail failure. The understanding of RCF in Europe has grown significantly since a tragic accident in Hatfield, England in October 2000 when a rail with RCF fractured into hundreds of pieces under a train travelling at speed and sadly 4 people lost their lives. [37] Potential solutions, other than rail management such as lubrication and rail grinding, include the use of innovative steel grades and modified alloys in the production of special rail steels. This has included the development of heat treated rail steels of hardness grades from 400 to 440 BHN to prolong the service life of rails on heavily used lines.

Fig. 2-24: An example of rolling contact fatigue gauge corner cracks [38]

The rails are installed on the sleeper with an inclination towards the track axis, which is meant to positively support the guidance of the vehicle. In principle, inclinations between 1/20 and 1/40 are differentiated in accordance with TSI Infrastructure. [8, section 4.2.4.7.1.] The rail inclination on the track has the following values in the DACH countries:

Germany   1/40 [39]

Austria   1/40 [19]

Switzerland   1/40 [7, AB 16(1.2.1)]

UK   1/20 [40]

Inclined rails are generally not installed in turnouts. However, due to the transition area required between inclined and straight rails, tests are currently under way to install inclined rails in turnouts as well. TSI Infrastructure specifies the inclination of the rail at speeds above 250 km/h. [8, section 4.2.4.7.2.] In the UK, prior to the introduction of 113A flat bottom rail circa 1970, rails were inclined in turnouts except for the switches. Today, rail inclination has returned in new turnouts with 60E rails.

To ensure the safe guidance of the vehicles, it is necessary to position rails at a defined distance from each other. This distance is called the track gauge. The following definition applies in accordance with EN 13848-1:

"Track gauge, G, is the smallest distance between lines perpendicular to the running surface intersecting each rail head profile at point P in a range from 0 to Zp below the running surface. In this standard Zp is always 14 mm." [41, section 4.2.1]

The European nominal gauge is 1435 mm. [8, section 4.2.4.1.]

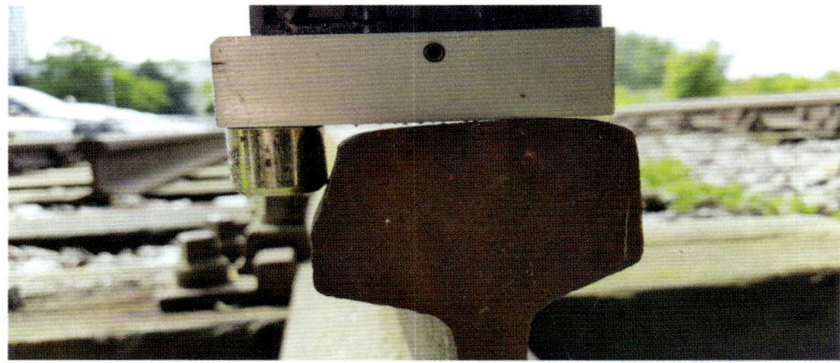

Fig. 2-25: Measurement of the gauge in a region of 0–14 mm below the running edge

Accordingly, the German EBO defines the primary gauge as 1435 mm, with the main line gauge not being permitted to be larger than 1465 mm. The lower limit of the gauge is generally limited to 1430 mm. [5, Art. 5(2) and (3)]

In the ÖBB network, the European nominal gauge is applied as the national nominal gauge. The basic dimension for the gauge is thus 1435 mm. [6, Art. 12(2)] To achieve better guidance, new track is laid with a design track gauge of 1437 mm. [42]

The nominal dimension for the design track gauge in Switzerland is generally 1435 mm. [7, AB 16(1.1.1)] In contrast to the existing standard, the gauge in Switzerland is measured exactly 14 mm below the connecting line at the highest point of the two rails. [7, AB 16(1.1)] Apart from a few exceptions as a result of historical developments, wooden sleepers are laid with a 46E1 rail profile with a gauge of 1437 mm (as second-hand material), and bi-block concrete sleepers are laid with a 54E2 rail with a gauge of 1433 mm. [14]

In the UK today, standard gauge is 1435 mm. measured 14 mm below the running surface. This was officially enacted through an Act of Parliament in 1846. Between 1964 and 1997 newly installed plain line and vertical S&C had a reduced gauge of 1432 mm, as it was considered at the time that this would provide a greater control of lateral wheel oscillation. Investigations into wheel/rail interaction in the late 1980's showed that this was not the case and it was decided to revert to 1435 mm, however, 1432 m gauge track could remain until renewed. [43]

The gauge may be widened on purpose during the installation of the track to provide more space for the wheelset in tight curves. This measure is an attempt to make it easier for the vehicles to travel through curves and to reduce wear.

DB		ÖBB		SBB		NR	
Radius [m]	Gauge Widening [mm]	Radius [m]	Gauge Widening [mm]	Radius [m]	Gauge Widening [mm]	Radius [m]	Gauge Widening [mm]
175 > R ≥ 150	1440	R ≥ 175	0	R ≥ 275	0	200 > R ≥ 176	3
150 > R ≥ 125	1445	175 > R ≥ 150	5	275 > R ≥ 185	6	175 > R ≥ 151	6
125 > R ≥ 100	1450	150 > R ≥ 125	10	185 > R ≥ 150	10	150 > R ≥ 126	9
		125 > R ≥ 100	15	150 > R ≥ 130	16	125 > R ≥ 101	12
				R < 130	20	R ≤ 100	15

Fig. 2-26: Comparison of gauge widening in tight curves in the DACH countries [14], [39], [44]

It is difficult to distinguish the profile and quality of rails with the naked eye. However, an unambiguous identification is crucial for the correct implementation of track construction and maintenance work. Therefore, EN 13674-1 specifies unambiguous rail markings by means of a rolling mark applied every four metres. [33] The rolling mark (Fig. 2-27) contains the mark of the manufacturing works (1), the steel type (2), the last two digits of the year of manufacture (3) and the designation of the rail profile (4 and 5).

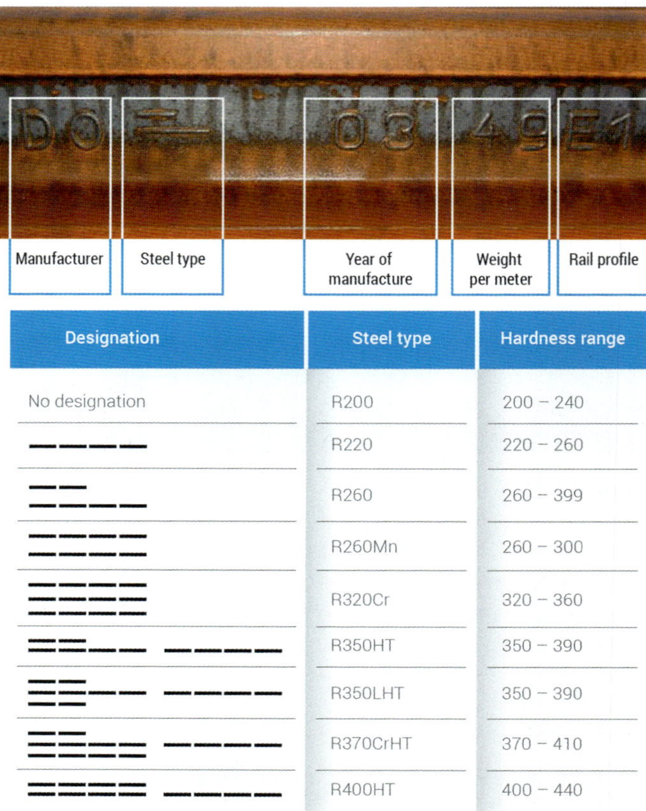

Designation	Steel type	Hardness range
No designation	R200	200 – 240
— — — — —	R220	220 – 260
— — / — — — —	R260	260 – 399
— — — — / — — — —	R260Mn	260 – 300
— — — — / — — — —	R320Cr	320 – 360
— — — / — — — — —	R350HT	350 – 390
— — / — — / — — — —	R350LHT	350 – 390
— — / — — — / — — — —	R370CrHT	370 – 410
— — — — / — — — —	R400HT	400 – 440

Fig. 2-27: **Example of the rolling mark (above) [45] and the bar code (below) for different rail grades as per EN 13674-1 [33]**

Rails are also used for other purposes. On critical structures, e.g. bridges or platforms, additional rails (known as guard rails) are installed on the sleepers to hold the train on the track in case of derailment. On bridges, guard rails are used in this way to either prevent the vehicle from falling off or to protect the structure from destruction. For this purpose, two rails are fixed between the two existing rails as a safety measure (see Fig. 2-28).

Guard rails are installed to prevent lineside structures from impact in case of derailment (e.g. bridge piers). They are fixed between the running rails on the side opposite to the object to be protected (see Fig. 2-29).

In curves with high cant deficiency, the wheelset of the vehicle is normally guided at the wheel flange along the outer rail of the curve. This forced guidance causes considerable wear on the rail. In curves with reduced radii, check rails may be installed to reduce the wear of the running rail. The check rail is installed on the inside of the inner rail of the curve for this purpose. In this way, the vehicle is guided on the back of the wheel flange on the inner check rail of the curve and not on the outer rail. The use of check rails is controversial [31, p. 73] and therefore, they are only deployed in very tight curves (in Austria for radii of less than 100 m [19]).

Disadvantages:

– Wooden sleepers are **susceptible to splitting**, and the sleeper has a **shorter service life** compared to other sleepers under similar loading.
– Their price is considerably higher than that of other sleepers.
– Wooden sleepers can **deform** and thus affect the track geometry noticeably. [31, p. 29]

The dimensions of sleepers can vary in turnouts and on bridges. Nails are driven into wooden sleepers through which the year of installation and the type of wood of the sleeper can be identified.

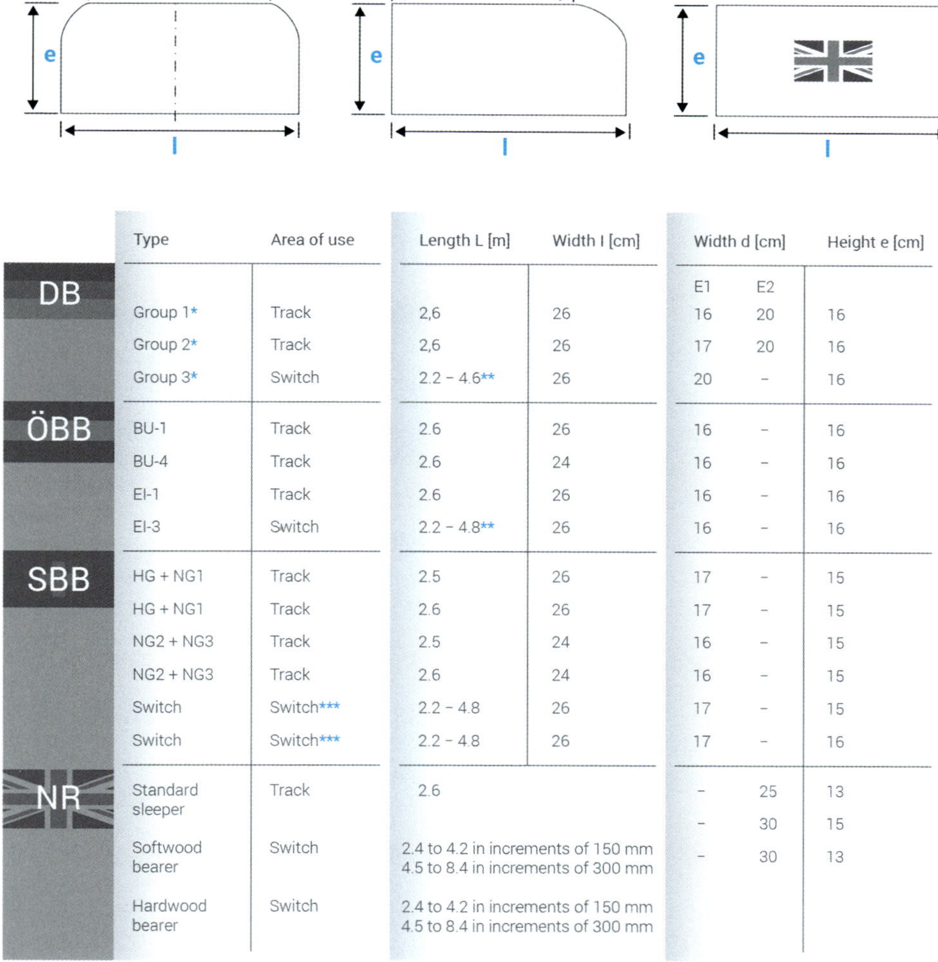

	Type	Area of use	Length L [m]	Width l [cm]	Width d [cm] E1	Width d [cm] E2	Height e [cm]
DB	Group 1*	Track	2,6	26	16	20	16
DB	Group 2*	Track	2,6	26	17	20	16
DB	Group 3*	Switch	2.2 – 4.6**	26	20	–	16
ÖBB	BU-1	Track	2.6	26	16	–	16
ÖBB	BU-4	Track	2.6	24	16	–	16
ÖBB	EI-1	Track	2.6	26	16	–	16
ÖBB	EI-3	Switch	2.2 – 4.8**	26	16	–	16
SBB	HG + NG1	Track	2.5	26	17	–	15
SBB	HG + NG1	Track	2.6	26	17	–	15
SBB	NG2 + NG3	Track	2.5	24	16	–	15
SBB	NG2 + NG3	Track	2.6	24	16	–	15
SBB	Switch	Switch***	2.2 – 4.8	26	17	–	15
SBB	Switch	Switch***	2.2 – 4.8	26	17	–	16
NR	Standard sleeper	Track	2.6		–	25	13
NR					–	30	15
NR	Softwood bearer	Switch	2.4 to 4.2 in increments of 150 mm 4.5 to 8.4 in increments of 300 mm		–	30	13
NR	Hardwood bearer	Switch	2.4 to 4.2 in increments of 150 mm 4.5 to 8.4 in increments of 300 mm				

* In Germany, the type of wood used is oak or beech.

** Exempt from this are excess lengths and the use as double sleeper.

*** In Switzerland, only oak is used for sleepers.

Fig. 2-34: Dimensions of newly installed wooden sleepers on track and in turnouts [14], [54], [55], [56]

2.6.2.2 Steel sleepers

In 1858, Le Crenier developed sleepers which were pressed out of a 4 mm sheet of steel. [31, p. 33] This advancement was driven mainly by the desire to develop a longer-lived sleeper material than wood. It was recognised early on that "in the whole world, only a fraction of that oak wood regrows that is placed under our rails every year to rot there within a few decades despite all preservation." [27, p. 127]

However, the first steel sleepers were not able to withstand the requirements placed on them. Over time, a suitable shape was developed, and the hollow steel sleeper has remained the same basic shape up to now. The standardisation of steel sleepers is less detailed compared to that of concrete and wooden sleepers (DIN 5904 [57] and [58]). In some cases, steel sleepers continue to be installed in the networks of the DACH countries, apart from Austria, where there is currently no provision for the new installation of steel sleepers. Although the geometric details differ between countries, the basic principle remains the same. This shape permits optimum positional stability due its position in the ballast bed, but care must be taken to ensure that ballast occupies the underside and that they are fully bedded in the ballast.

Tamping of track with hollow steel sleepers presents a particular challenge due to their shape (see Chapter 5). Initial problems with attaching the rail to the sleeper were largely resolved by welding on grooved plates during manufacture. [31, p. 35]

Fig. 2-35: Hollow steel sleepers (left) and Y-sleepers (right)

Advantages:
- The steel sleeper has **high positional stability** if bedded sufficiently into the ballast. [59, p. 1]
- Their low weight permits **easy handling**.
- Steel as a material is robust and guarantees a **long service life**, provided the sleepers are protected from corrosion.
- **Re-use** is possible through refurbishment of the sleeper. [31, p. 36]

Disadvantages:
- Steel is a very good conductor which leads to **insulation problems** with regard to the signal current.
- If insufficiently bedded into the ballast, their **low weight** has a negative effect on the positional stability of the track.
- **High purchase cost** contrasts with the reduced number of sleepers required per track kilometre.
- Severe deformation in case of derailments cannot be prevented. [31, p. 36]
- Steel sleepers with long ends can lead to problems during **lining** by a tamping machine due to their high lateral resistance.

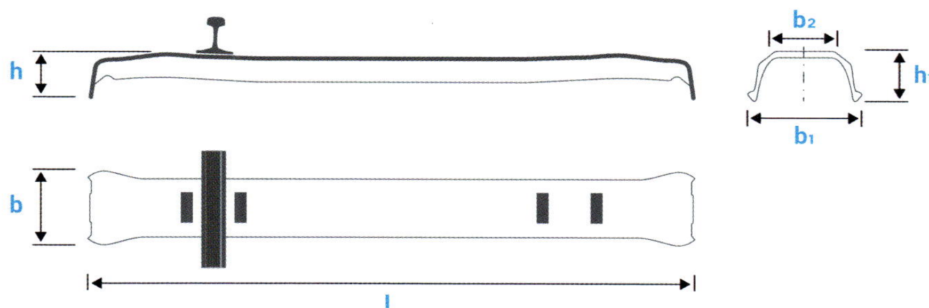

Fig. 2-36: Schematic diagram of a steel sleeper

	Type	Length L [m]	Width b [mm]	Width b_1 [mm]	Width b_2 [mm]	Height h end plate [mm]	Height h_1 [mm]
DB	St 82	2.4 / 2.5 / 2.6	365	260	145	150	100
ÖBB	Sw 9	2.5	365	260	135	150	100
SBB	46W1 54E2	2.556 2.575	320 320	240 240	130 130	145 170 (237)*	90 90
NR	British Steel Section 436	2.580	260	260	168	100	100

* For new sleepers, height h of the end plate has been increased to 237 mm.

Fig. 2-37: Typical dimensions of steel sleepers [10], [14], [60], [61]

Over time, another type of sleeper has been developed from the hollow sleeper: the Y-sleeper. It takes its name from its characteristic shape (Fig. 2-35 (right)). Two curved I-beams (the I refers to the characteristic cross-section of the beam) connect six rail support points (three double bearing points). The two steel beams are welded together with steel flanges at the top and bottom.

Due to this special configuration, track with Y-sleepers has a higher lateral displacement resistance than track with concrete sleepers despite their comparatively low weight of about 143 kg. [59, p. 21] In tight conditions, the overall construction height can be reduced due to the engagement of the sleeper with the ballast. Moreover, Y-sleepers, with a length of 2.30 m, have smaller dimensions so that they are of particular interest for a ballast bed of low width. The track geometry must be corrected with special tamping machines on which the tamping bank can be adjusted for this special sleeper type. Despite some advantages, the Y-sleeper has not caught on due to the higher unit cost and the special maintenance requirements.

Even so, it can be found at particularly stressed points with low ballast bed thickness, small radii or in stations. In Germany, the use of Y-sleepers is possible up to speeds of 120 km/h on ballasted track. In Switzerland, too, steel Y-sleepers are installed, but they are not approved for all applications.

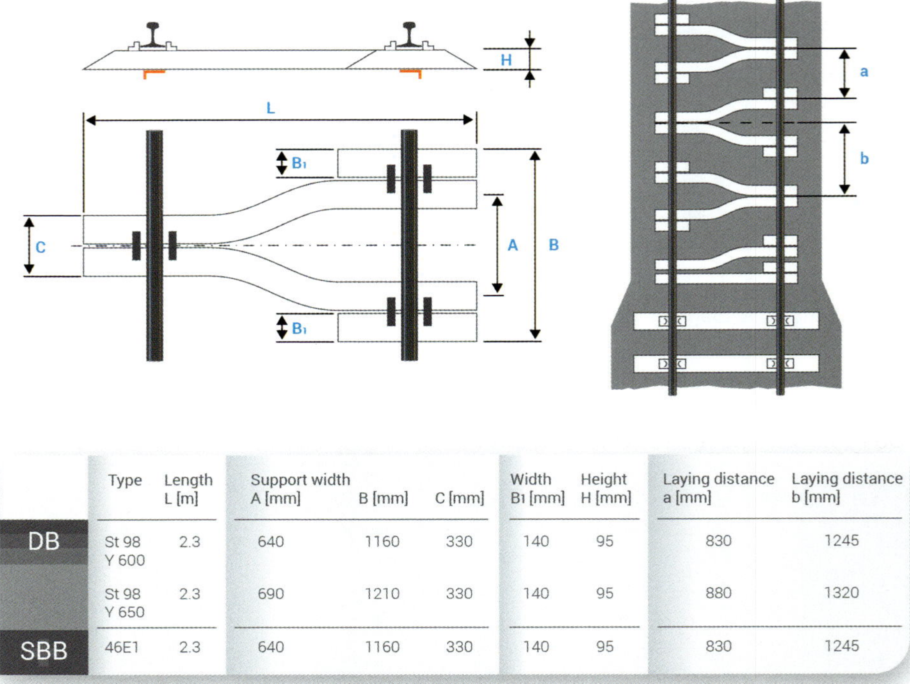

	Type	Length L [m]	Support width A [mm]	B [mm]	C [mm]	Width B₁ [mm]	Height H [mm]	Laying distance a [mm]	Laying distance b [mm]
DB	St 98 Y 600	2.3	640	1160	330	140	95	830	1245
	St 98 Y 650	2.3	690	1210	330	140	95	880	1320
SBB	46E1	2.3	640	1160	330	140	95	830	1245

Fig. 2-38: Typical dimensions of Y-sleepers [14], [62], [63]

2.6.2.3 Concrete sleepers

The stresses on sleepers are varied. In addition to classic compressive stresses, tensile stresses also occur due to wheel forces and the tilting movement of the rail. These forces act on each sleeper and are of particular interest concerning the historical development of the concrete sleeper. Concrete absorbs compressive stresses very well, depending on its compressive strength, but tears under tension. This was recognised by the French gardener Joseph Monier, who designed the first concrete sleeper in 1884. [64] Monier used non-prestressed reinforced concrete at that time to absorb the tensile stresses, but the newly developed concrete sleepers did not withstand the loading.

Concrete is very good at absorbing and transferring pressure. However, under tensile stress it starts to crack and tends to sudden failure in case of overload. This behaviour originally restricted its use in construction work. To improve the characteristics of concrete, it was combined with steel. In contrast to concrete, steel can absorb tensile stresses due to its material properties and does not fail suddenly but is characterised by its ductile behaviour. To make concrete components for construction, steel in the form of rods, the so-called reinforcing steel, is installed in the zone of the concrete that is subject to tensile forces. This is the area in which the material expands due to stress. As a combination of both materials, reinforced concrete merges their properties.

Only after World War 2, with the advent of prestressed concrete sleepers, a real alternative to the then predominant wooden sleepers was created. In prestressed concrete sleepers, the reinforcing steel inserted into the sleeper is prestressed. For this, the steel is stretched on purpose to a specified tension. There are various methods for stretching, but the main effect is the same. Prestressing the steel generates compressive stresses in the sleeper prior to installation. This counteracts the externally applied loads after installation.

Advantages:

- As a sleeper material, concrete has a **long service life**.
- The high weight of concrete sleepers has a positive effect on the **stability** of the track, provided the subsoil has sufficient load capacity.
- **Low production costs** reduce the price of concrete sleepers.
- Concrete is mostly weather-resistant and resistant to other environmental effects.

Disadvantages:

- Concrete sleepers are sensitive to **impact**.
- Manual laying and manipulation, due to the high weight, are only possible with difficulty.
- A lower overall elasticity leads to higher loads on the ballast.
- If the subgrade has a low load capacity, the characteristics of the concrete sleeper have a negative effect on the development of the track geometry.

The first concrete sleepers copied the shape of wooden sleepers, but over the years, different profiles developed due to the easy shaping of concrete. In addition to the **monobloc concrete sleepers** mainly used in the DACH countries, the **twin-block concrete sleepers**, frequently used in France, were also developed.

For a twin-block sleeper (see Fig. 2-39), two concrete blocks which serve as rail support are joined by a rolled steel tie bar. The reinforcement of the concrete blocks is usually non-prestressed as the tensile stresses are absorbed by the steel tie bar. Despite their low weight, twin-block sleepers have a high lateral displacement resistance but are particularly susceptible to twisting and damage during maintenance due to their shape.

Fig. 2-39: Twin-block sleepers in Switzerland

In addition to the usual shapes of monobloc sleepers (see Fig. 2-40), there are also some special shapes, e.g. the HD sleeper.

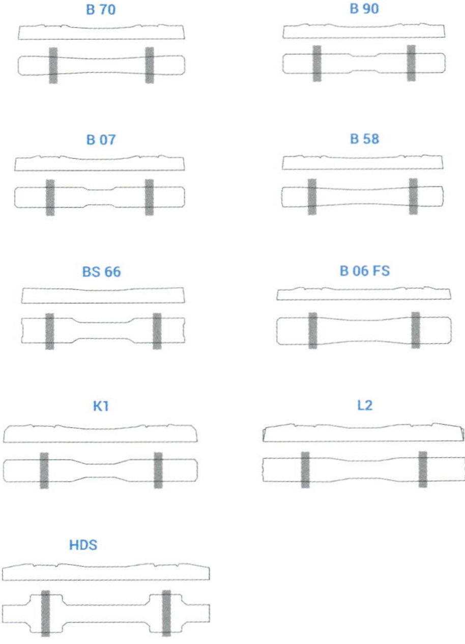

Fig. 2-40: Common concrete sleepers in the DACH countries [14], [65] – [69]

	Type	Length L [m]	Sleeper depth at rail seat H_A [mm]	Sleeper height at the centre H_B [mm]	Support width at rail seat B_A [mm]	Sleeper weight* [kg]
DB	B 70 W	2.60	210	175	300	304
	B 90 W	2.60	200	175	320	355
	B 07	2.60	200	175	320	360
	B 58 W	2.40	200.5	175	300	249
	BS 66	2.27	217	177.5	350	276**
	B 06 FS	2.60	172	153	400	361
ÖBB	K1	2.60	220	180	300	282**
	L2	2.6	215	185	300	291**
	HDS	2.6	233	198	550	250
SBB	B 70	2.60	210	175	300	290
	B 91	2.60	210	175	300	286
NR	G44	2.50	200	175	200	312

* The weight specifications are guidelines only and refer to the sleeper including the common fastening material

** Without rail fastening

Fig. 2-41: Common dimensions of monobloc sleepers in the DACH countries and the UK [14], [65] – [69], [70]

2.6.2.4   Under-sleeper pads

The different mechanical properties of concrete sleepers compared to wooden sleepers result not only in a noticeably lower elasticity of the overall system but also lead to a reduced contact surface with the ballast due to the lower engagement. [71, p. 16] This means that wheel loads are transferred across fewer sleepers (lower deflection of the system) and through a lower number of ballast stones into the substructure. The load on the ballast is therefore imparted through fewer points of contact, making the spread of the load more concentrated. This fact is reflected in the rules and standards of the railways. For concrete sleepers, these frequently demand a separate examination that the load capacity of the system is sufficient. The properties of concrete sleepers and the ballast bed can be improved in the long term by using under-sleeper pads directly below the sleeper to increase the degree of vertical elasticity in the track superstructure. [72, p. 67] Under-sleeper pads are either introduced directly into the still wet concrete during sleeper production or are glued to the underside of the sleeper after production. The pads consist of various highly specialised plastics.

Depending on the material used (elastomer or thermoplastic materials), the under-sleeper pads reduce wear on the track system through lowering dynamic forces, thereby improving the durability and cost-efficiency of the overall system. They have a thickness of 5 to 20 mm. Usually, the perimeter of under-sleeper pads is 5 to 15 mm smaller than the underside of the concrete sleeper to prevent damage, if any, from tamping tines. [73, p. 8]

2.6.3   Track fastenings

This track component is of more importance than one might assume at first glance. We will use the following definition in this book:

"The totality of special elements which connect the rail to the sleeper or other supports and the rails to each other". [31, p. 51]

Thus, the following components of the track are covered by the term track fastenings in this book:

– rail fastening system,
– rail joints and
– anti-creep devices and sleeper anchors

2.6.3.1   Rail fastening system

The purpose of rail fasteners is to attach the rail to the sleeper or other supports, such as bridges, in the position required. The rail fastening system is subject to extreme stresses. The forces applied must be transferred to the sleeper without changing the gauge between the two rails. Horizontal shifting of the rail and its tilting, either inwards or outwards, must be prevented. The load distribution from the rail to the sleeper is managed by the provision of elastic elements (rail pads). These compensate for the impact loading and thus provide support to the overall system. Depending on the design, modern fasteners consist of mounting elements (screws, spikes or sleeper clips) and load-distributing and elastic elements (baseplates and rail pads).

Worldwide, a variety of different fastening systems have developed which differ with regard to their characteristics, such as function, holding-down force and spring characteristics. They can be combined into two main groups by design:

- The **direct rail fastener** fixes the rail and the baseplate on the sleeper with the same mounting element. If the mounting element is released, the baseplate and the rail are released from the sleeper.
- The **indirect rail fastener** uses a two-part mounting system whereby the baseplate is attached to the sleeper through its mounting system separately from the rail.

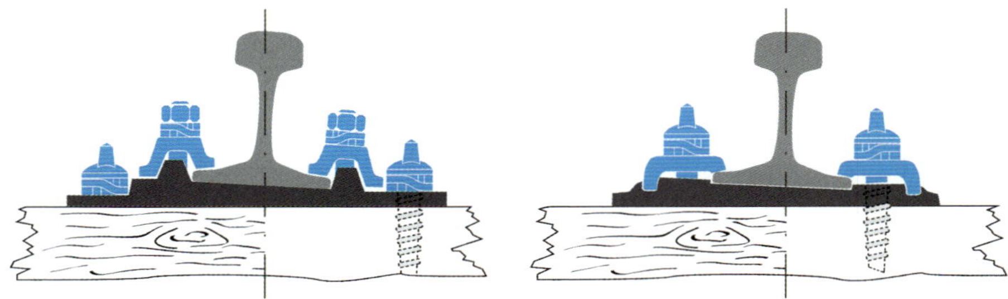

Fig. 2-42: Indirect (left) and direct (right) rail fastening systems

In addition to the direct bolting of steel sleepers, two systems have mainly prevailed in the DACH countries: the indirect clamping plate system (K-type track) and the direct angled clamping plate system (W-type track).

Fig. 2-43: UK rail fastenings; Top: Bull Head chairs and elastic spikes.
Bottom: Pandrol Clip and the Pandrol Fastclip [74]

The clamping plate system (Fig. 2-44 left) utilises the clamping plate to attach the rail to the ribbed baseplate underneath. The longitudinal ribs of the ribbed baseplate, which lie in the direction of the rail, support the gauge uniformity of the rail. A rail pad is placed between rail foot and ribbed baseplate to dampen the contact forces and increase the twisting and creep resistance of the fasteners.

In the UK, with the predominate use of bull head rail until 1948, the fastening system was a cast iron chair with a wooden key, acting as a wedge, that secured the rail into the chair and maintained gauge. The adoption of flat bottom rails led initially to the use of baseplates and elastic spikes to secure the foot of the rail to the plate and the plate to the wooden sleeper (see Fig. 2-43 top right). The growing use of concrete sleepers led to British Railways adopting the Pandrol clip, invented by Norwegian Per Pande-Rolfsen in 1957, as standard in 1966. Whilst there have been many developments of the clip over the years, particularly the e-Clip, the currently approved product for 200 kph track is the Fastclip FC.

The Pandrol Fastclip was first introduced onto the UK main line rail network in 2000. It was the first clip to be pre-installed onto the sleeper before being brought to site and is today the standard fastening for new plain line concrete sleepers.

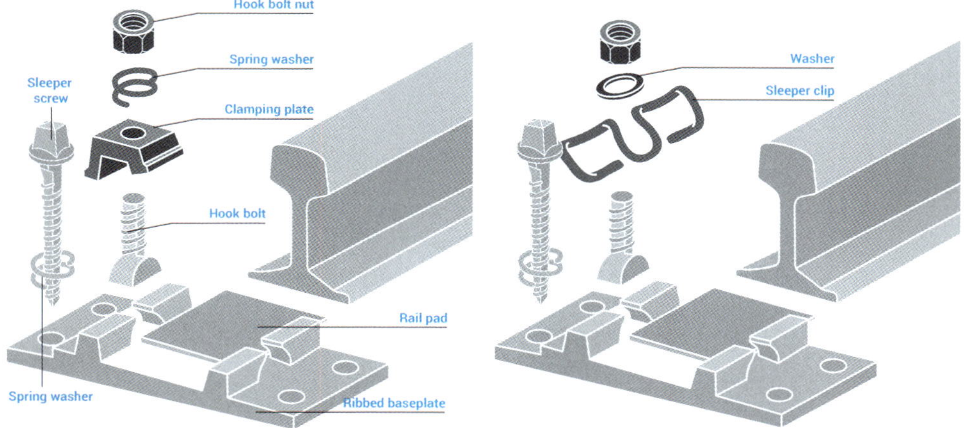

Fig. 2-44: Schematic diagram of a clamping plate system (left: K-type track) and a clip system (right: KS-type track)

The need and benefit of providing a resilient pad between the underside of the rail and the fastening or sleeper has been known for many years. Originally, rail pads of pressed poplar wood were used. Nowadays, elastic materials (e.g. plastic or rubber) are used, which also have a positive effect on the load bearing performance of the system. For the adapted clamping plate system (Fig. 2-44 (right)), the clamping plate is replaced by a rail clip. However, the basic manner of operation remains comparable.

Baseplates are not required for concrete sleepers due to their particular shape in the area of the rail support. The rail sits directly on the sleeper, separated only by the elastic rail pad. The rail is fixed in position utilising angled clamping plates at the side and the sleeper bolts (Fig. 2-45).

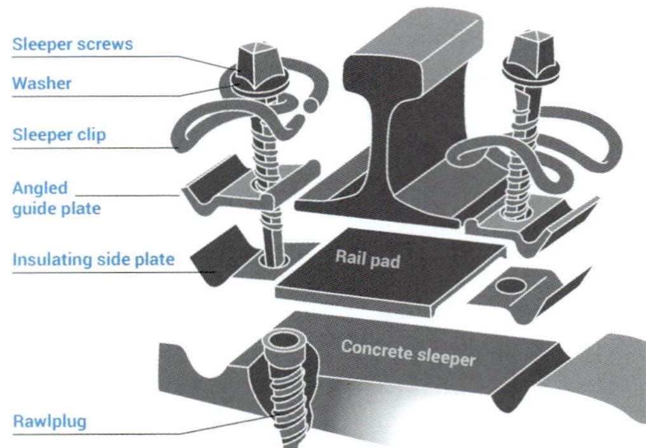

Fig. 2-45: Schematic diagram of an angled clamping plate system (W-type track)

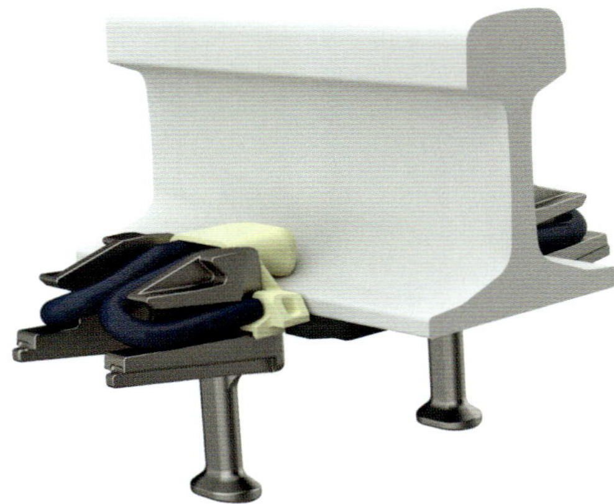

Fig. 2-46 The Pandrol Fastclip FC rail fastening which is the standard used in the UK on main routes [75]

If the load transmission between rail and sleeper is no longer sufficient (loss of holding-down force), excessive stresses arise in the individual components. This is reflected by their wear (e.g. rail faults or sleeper fracture). In extreme cases, it is no longer possible to lift and line the track panel as required during tamping. As rails and sleepers are no longer connected, the sleepers stay in the ballast during lifting. It is, therefore, sensible to check the proper functioning of the fasteners before a tamping activity.

2.6.3.2 Rail joints

Rails are supplied to the construction site in the required rolled length. On site, the rails can be joined together either by means of releasable (bolted) or fixed (glued insulated or welded) rail joints. [31, p. 76]

With the development of advanced welding technologies, railways have, over the last 50 years, changed over to producing a continuous welded track. Welding of the two rail ends has clear advantages due to the formation of a continuous rail surface.

The gap, even if small, between the rail ends (joint gap) of releasable rail joints inevitably increases the stress on the overall system. These large forces damage individual components in the longer term, reduce the ride quality, increase noise emissions and therefore in modern railways releasable rail joints are only used on branch lines or in tight radii.

Fig. 2-47: Supported rail joint on wooden sleepers (left) and suspended rail joint on concrete sleepers (right)

Despite all attempts, joint gaps in rails cannot be completely avoided. Where the electric conductivity of rails is used for track circuits in signalling, it is necessary for the track sections to be electrically isolated. Therefore, insulated joints are used at these locations which electrically insulate the track sections and divide the line into sections. [76, p. 26]

For jointed track, especially in the DACH countries, depending on the position of the joint with regard to the joint sleepers (Fig. 2-47), a distinction is made between supported rail joints (twin sleeper joints) and suspended rail joints in the centre of the crib. The two rails are connected with fishplates on both sides of the web of the rail. The web is predrilled depending on the fishplate used. Once the joint has been set up, the fishplates can be mounted on both sides of the web using fishplate bolts, and through torqueing the bolts, the transfer of longitudinal forces is applied.

Not only are jointed rails particularly costly, but they also place special requirements on maintenance. When correcting the track geometry, particularly large correction values are to be expected in these areas due to the high level of wear of the components. Furthermore, when tamping the track, particular care must be taken with the twin sleepers and the setting of the tamping tines. It should be noted that the practice of double sleepers at rail joints is not used in the UK.

Fishplates may also be used as emergency fishplates when repairing a broken rail. However, this is only an interim solution as a rule.

2.5.3.3 *Anti-creep clips and sleeper anchors*

Anti-creep clips (rail clips), sleeper anchors and sleeper end plates have the aim of stabilising the lateral and longitudinal position of the track, particularly on curves.

In general, two different systems of anti-creep clips – "Fair" and "Mathée" – are used in the DACH countries. They are used where a longitudinal movement of the rails is to be prevented (e.g. due to temperature changes). In both systems (see Fig. 2-48), a clamping device supported against the sleeper is fitted to the rail and thus makes a longitudinal movement more difficult. If a change in length of the rail is required during maintenance, the anti-creep clips must be removed first. Similar rail anchors were used on flat bottom jointed tracks in the UK.

Fig. 2-48: Anti-creep clip "Fair" used on concrete sleepers

Sleeper anchors and sleeper end plates increase the lateral resistance of the track. They are fitted either to the centre of the sleeper (sleeper anchor) or the sleeper ends (sleeper end plate) on the inside of the curve and are used to significantly reduce the risk of buckling. However, they are also used at platforms, in turnouts and to improve the horizontal stability of the track. They must be released for tamping the track and refitted afterwards. Similar rail anchors are used, especially on sharp curves, in the UK.

Fig. 2-49: Sleeper end plates in a tight curve [77]

2.6.4   Track codes

The permanent way has numerous components. Their combination defines the characteristics of the track and its effect as a system. A simple representation of the components using codes makes it easier, and also quicker, to assess the situation on site. In this way, a large quantity of information can be displayed comparatively simply by means of various abbreviations which are meant to facilitate quick naming. These codes are described in Germany as 'track arrangement' [78] and in Austria as an abbreviation of the track type [79].

In Germany, the continuous welded track is generally designated as follows:

Rail fasteners – rail type – sleeper type – number of sleepers

W: rail fastened with sleeper bolts or rail clip

K: rail fastened with hooked bolt and clamping plate

KS: rail fastened with hooked bolt and rail clip

The above letters are always followed by the exact designation of the fastener including the rail pad used. The rail profile used is given, as usual, by the weight per linear metre and separated by a hyphen from the rail fastener. Sleeper types are distinguished as follows:

H: wooden sleepers

St: steel sleepers and

B: concrete sleepers

This letter is combined with the precise sleeper designation. Finally, the number of sleepers per 1000 m is stated. Example:

W 14K – 60 – B 70 – 1667

This is an angled guide plate with rail clip (or hooked bolt) of type Wfp 14K which fixes a rail of rail profile 60E2 to a concrete sleeper of type B 70 W with a spacing of 60 cm.

In case of a jointed rail, the above code is extended by the track arrangement. The information on the number of sleepers is replaced by:

$$\frac{\text{Type of rail joint + number of sleepers}}{\text{Rail length (length of track panel)}} - \text{Sleeper spacing}$$

This results in, for example:

$$\text{KS} - 54 - \text{H Gr. 1} - \frac{\text{Br} + 99}{60} - 60$$

In Austria, the abbreviation of the track type contains more information:

Rail profile – rail length – sleeper shape – rail fasteners – sleeper spacing

Furthermore, it can contain information on the type of under-sleeper pads and any gluing of grooved plates. For example:

60E1 – Lv – Be K1 – Op (Skl 14, Wfp 14 K, Zw K2a) – 600

In this case, this is a longitudinally welded 60E1 rail profile, which has been fixed with rail clips of type Skl 14, angled guide plate 14 and rail pad Zw K2a on concrete sleepers of type Be K1 with a spacing of 600 mm.

49E1 – 30 – Bu1 – Rp+Rp – 650

This is a jointed track with a rail profile 49E1, a rail length of 30 m and beech twin sleepers with grooved plate fastening. The sleeper spacing is 650 mm.

Due to the variety of track components, it is only possible to give a few examples of the different codes. Further information is available in the rules and standards of the individual railways.

In Switzerland, abbreviations [14] are also used for the different track components; however, a combination used as code is not customary, as far as we know. Nor is there a track code in the UK.

2.6.5   Ballast

The purpose of railway ballast is to transfer and distribute the loading from the sleepers via the substructure and into the subsoil. In so doing it ensures a stable position of the track panel in the horizontal and vertical direction. The subsoil is protected by the distribution of forces through the ballast bed. The ballast bed is no indiscriminate jumble of large and small stones but has a tailored composition by size and rock type. Ballast stones of different sizes follow a defined distribution and thus form different percentages of the overall structure. Using square mesh screens of appropriate hole sizes [80], the ballast suppliers crush the quarried rock and mix the ballast composition required from individual grades according to a "recipe".

The size selected is stated as a combination between the smallest and largest size. In Switzerland, ballast of group A (31.5/50) is used, while in Germany and Austria mainly ballast of size group B (31.5/63) is used.

Deviating from this, ballast of size 16/31.5 is also used in Austria. This ballast is used in areas which are stepped on by staff more frequently in the course of their work (e.g. shunting). However, excepted from this are sections with a loading $\geq$ 20,000 total gross tonnes or a speed of $v_{max} \geq$ 80 km/h. [19]

Network Rail's specification for ballast is defined in their standard NR/L2/TRK/8100 and meets the requirements of the European size group C.

Screen size	Screenings, per cent by weight				
mm	A	B	C	D	E
80	100	100	100	100	–
63	100	95 to 100	95 to 100	93 to 100	–
50	70 to 99	65 to 99	55 to 99	45 to 70	100
40	30 to 65	30 to 65	25 to 75	15 to 40	90 to 100
31.5	1 to 25	1 to 25	1 to 25	0 to 7	60 to 98
22.4	0 to 3	0 to 3	0 to 3	0 to 7	15 to 60
16	–	–	–	–	0 to 15
8	–	–	–	–	0 to 2

Fig. 2-50: Ballast stone size distribution as per EN 13450 [80]

For the very early colliery railways in the north east of England, ballast was simply the gravel placed in coal ships returning there and unloaded at the quayside upon their arrival. From there it found its way into the colliery tramways and railways as a suitable material to keep the rails level. It is only in the last 50 years and the development of modern CWR track and high-speed lines that science has played a role in defining the composition of the ballast bed. Prior to that, all types of relatively free draining materials have been used, including ash from steam engines and broken slag from steel works. Track geometry was maintained by manually packing more of the same material, or fine stone chippings, under the sleepers. The development of tamping machines has also influenced the specification for specific geometric, physical and chemical requirements of ballast that is today defined in EN 13450. [80] Worldwide, crushed stone has largely prevailed as the material of choice for railway ballast.

Over time, different types of railway track developed, e.g. the non-ballasted track, which is completely without ballast. Nonetheless, the ballasted track has proven itself and adapted to the requirements of modern railway operations. Due to the defined size distribution and sharp edges, individual ballast stones interlock and thus form a stable, but elastic free draining support grid for the track panel. In addition to size distribution, the stone shape (e.g. flatness and length) is also important for the success of the system and is described in the standard in more detail.

The loads applied lead to a rearrangement of the ballast stones and also to degradation of the ballast, which consists mainly of the following individual phenomena:

- **Stone breakage**
  When an impact load is applied, the individual ballast stones break, and the size distribution changes. Different loads can be withstood depending on the material used. Clearly defined physical requirements (e.g. LA coefficient or micro-Deval coefficient) attempt to define the quality as well as possible. These parameters largely depend on the stone material used. In the DACH countries, granite, gneiss and diabase are used, whereas in the UK it is primarily granite. Compared to limestone, these types of rocks have a much higher resistance. Clearly, if the stones break faster, the life of the ballast bed will be markedly reduced.

- **Rounding**
  Over time, the loading changes the shape of the ballast stones. Sharp edges are chipped away and the interlocking mechanism of the stones as a grid is reduced. Consequently the ballast has difficulty in keeping the track in a stable position.

- **Attrition**
  The ballast stones rub against each other, fines are produced, resulting in abraded particles clogging the stones over time. Once a certain percentage is reached, not only is the load capacity reduced, but also the ability of the ballast bed to freely drain water is inhibited. A wet ballast bed increases the rate at which the stones wear (wet attrition) and in winter, ice lenses can form leading to an unintentional settlement when the temperature rises and the ice thaws.

Ice lenses arise from the freezing of water in the track bed. The volume of the water increases, and the remaining ballast stones are displaced. The resulting frost heave leads to a lifting of the track panel which must be levelled out. In the spring, the ice melts and a void forms. The ballast bed then suddenly gives way under the load of a train travelling over it and the resulting settlement leads to a dramatic worsening of the track geometry.

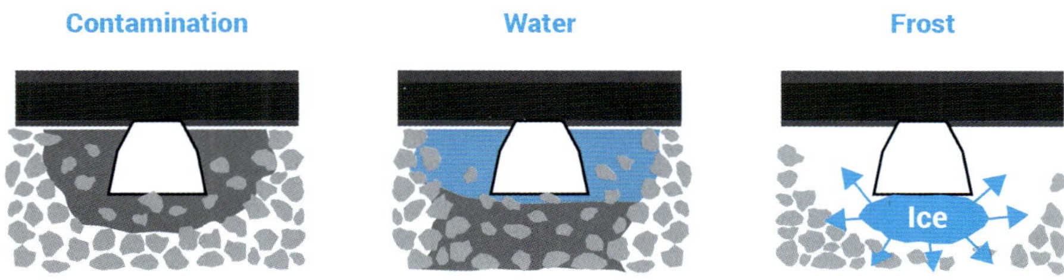

Fig. 2-51: Effect of an ice lens on track geometry, based on [31, p. 96]

A consistent standard of track geometry is largely dependent on the quality of the ballast material, the load applied, and a uniform depth of ballast bed. The continuous loading from trains leads to permanent (plastic) irregular deformation of the ballast bed. Any change in longitudinal uniformity of the ballast bed, for example changes in its depth or bearing capacity, is transferred to the longitudinal profile of the track panel. This is due to the plastic deformation of the ballast bed remaining when the track is unloaded, unlike the elastic properties of the track panel. This causes unintended changes in the track geometry. This not only has an initial negative effect on the ride quality, but also sets in train a gradual deterioration of the track geometry, similar to a small pothole in the road. Fortunately, it is possible to correct these defects and reinstate the original track geometry or very close to it, using various maintenance techniques.

A vehicle can compensate for irregularities in the track, depending on the speed at which it travels, due to the design of its suspension system. However, if the track defect exceeds defined threshold values, safe travel on the track is no longer possible and the permissible speed must be reduced. Any speed restrictions that may be required affect the timetable and reduce the availability of the line, thereby causing delays. Infrastructure managers set different track quality maintenance intervention levels depending on the speed of the line (see Chapter 4).

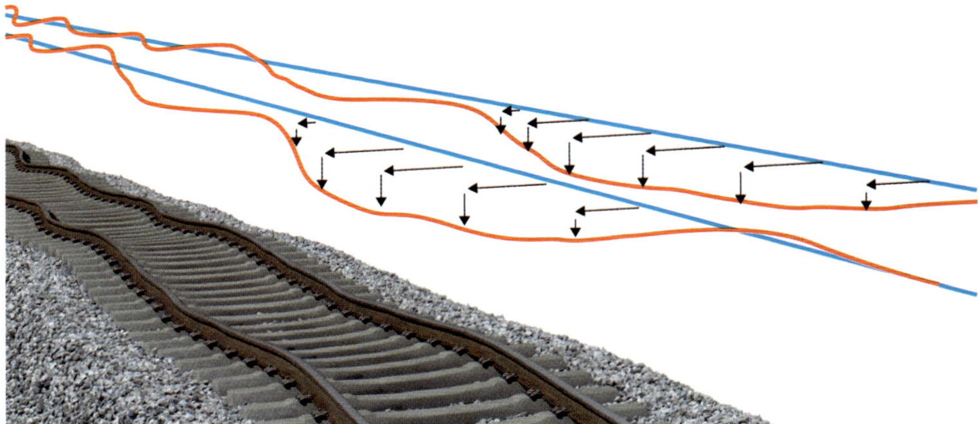

Fig. 2-52: Plastic deformation leads to a change in the track geometry

Over time, the composition of the ballast bed, and thus its properties, changes due to the wear described above. Newly installed ballast consists of the sizes specified in the size distribution and, in the first place, is made up of stones of regular shape (see Fig. 2-53). Wedge-shaped grains (15-30 mm) follow from initial stresses, which at first have a positive effect on the properties of the ballast. A certain percentage of small grains supports the load capacity of the ballast bed and is advantageous for the development of track geometry. The increasing share of fines (filler grains) fills the voids in the ballast more and more and starts to surround the wedge-shaped (spacer) grains. This, however, gradually changes the properties for the worse. To be able to react to this change in ballast properties, it is possible to adjust the individual parameters of the tamping machine to the track bed condition.

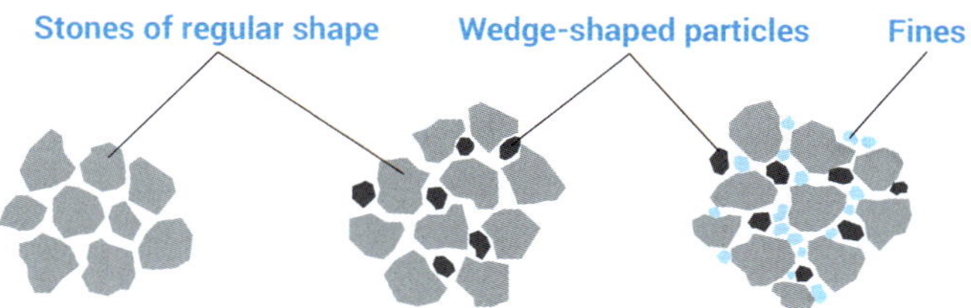

Fig. 2-53: Different phases of ballast wear, based on [31, p. 95]

Due to the above wear phenomena, the rounded ballast stones lose their ability to engage so that, in combination with the fine grain arising in the ballast bed, it becomes increasingly difficult to achieve a stable track geometry. The elasticity of the ballast bed is reduced because old and contaminated ballast can no longer dampen the impact. The track panel must now take up this additional stress which frequently results in component damage and fractures.

In addition to the load capacity, the ballast bed must also ensure sufficient drainage capability (permeability). If water remains in the ballast bed, this weakens the load capacity of the system with a lasting effect. It acts between the stones like oil on a bicycle chain.

Although the ballast bed has to be regarded as a structure, high demands are made on individual stones. The type of stone used must not only have appropriate resistance to breaking and to impact loads but also withstand weather effects, such as frost and solar radiation.

Initial wear of the ballast bed in the early years of its life can be managed by good maintenance of lineside drainage and correction of defects in track geometry. As the ballast bed continues to show signs of wear, indicated by the reducing intervals for action to correct the geometry and the poorer quality of measured geometry, a decision will be taken to renew the ballast. The infrastructure manager has options, depending on budget and track condition, to renew the track completely, renew the ballast only, or, by using modern mechanical machinery, to clean the ballast only. This latter process removes all the fines and soil before returning cleaned old stones mixed with new. It is also possible to re-sharpen the old stones as they pass through the machine.

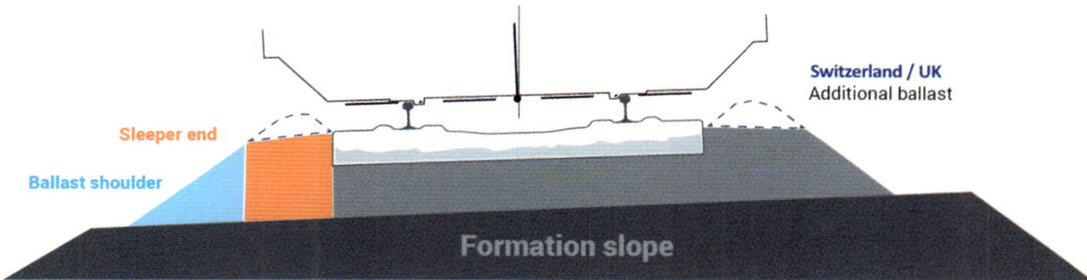

Fig. 2-54: Diagram of the individual components of the ballast cross-section in the DACH countries and the UK

In addition to its material properties, the shape of the ballast bed also has clear technical objectives. These are load distribution, sufficient lateral resistance and also ensuring free drainage. If one looks at a cross-section of the ballast bed (see Fig. 2-54), it looks like a trapezoid. The track panel sits on top of the ballast and is fully surrounded by the ballast. This applies to the sleeper crib between the sleepers and the ballast shoulder. In Switzerland and in the UK, the quantity of ballast at the shoulder is increased to improve the properties of the ballast bed, in particular concerning lateral resistance. [61]

In Austria and Germany, the ballast normally goes up to the upper edge of the sleepers. At very high speeds, suction forces arise under the trains due to the air turbulence. These forces can be so great that ballast stones are flung in the air and thus can damage buildings and trains in the vicinity or, in the worst case, endanger lives. This phenomenon is called flying ballast and must be avoided at all cost. [81, pp. 37f.] In Switzerland, the ballast is therefore generally lowered in the crib next to the sleeper. In Germany and Austria, this is also done, depending on the speed.

The ballast shoulder at the edge of the ballast bed has a specified angle. Ballast, as bulk material, cannot be placed at a steep angle without additional construction measures. It is, therefore, necessary to flatten off the profile of the ballast bed at the edges.

A specified thickness of ballast has to be maintained underneath the sleeper. If the thickness is too low, the system reacts stiffly to the forces applied and wears particularly quickly. Too much ballast, on the other hand, leads to high batter, inhomogeneity and high stresses on the rail. Therefore, the maximum thickness of the ballast bed is generally limited to 70 cm in the DACH countries to ensure homogeneous support of the track panel. In a similar manner to the suspension of a car, a uniform vertical stress distribution from the wheel to the ground is crucial. Homogeneity across the complete line is of particular importance here.

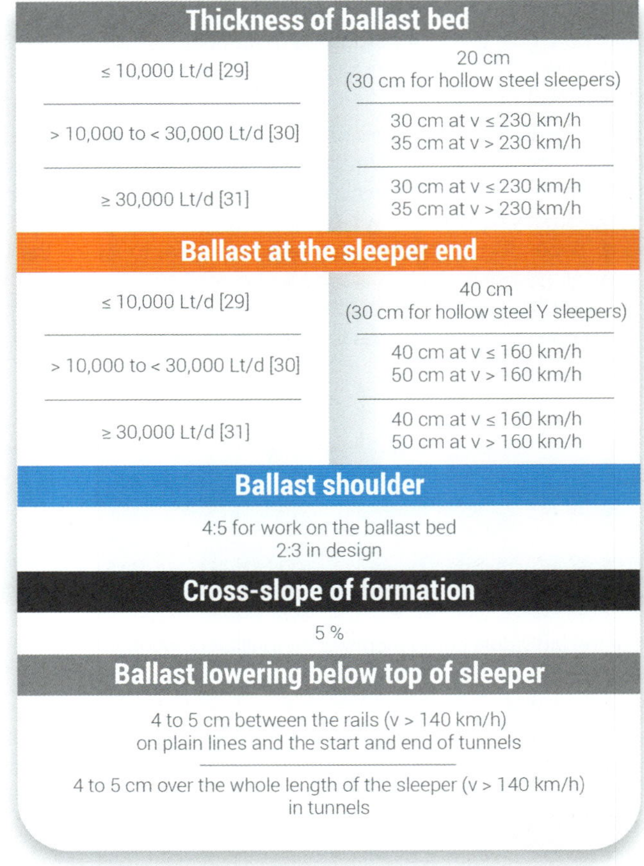

Fig. 2-55: Data on the ballast bed cross-section for plain line track in the DB network [16] – [18], [82], [83]

In Germany, the thickness of the ballast bed is measured from the underside of the sleeper of the non-canted rail of the track. [12] For steel sleepers, the thickness of the ballast bed is measured from below the top panel of the sleeper. Depending on the track load and the speed, this results in a required thickness of the ballast bed, as shown in Fig. 2-55.

The figure only shows standard cases; the guideline has to be consulted for exceptions.

In contrast to Germany, the thickness of the ballast bed in Austria is measured from the top of the sleeper and not from the bottom of the sleeper. It, therefore, depends on the type of sleeper used.

Thickness of ballast bed

S	Concrete with pads	50 cm
1	Concrete with pads	50 cm
1, 2	Concrete	50 cm
3, 3G	Concrete	45 cm
S, 1	Wood	45 cm
2	Wood	35 cm
3,3	Wood	30 cm

Ballast at sleeper end

S	Concrete with pads	55 cm
1	Concrete with pads	50 cm
1, 2	Concrete	50 cm
3, 3G	Concrete	50 cm
S, 1	Wood	50 cm
2	Wood	50 cm
3,3	Wood	45 cm

Ballast shoulder

Ballast shoulders 2:3 (4:5 also on existing tracks)
If the refuge is not specifically marked
in the cross section (e.g. cable chute),
the ballast shoulder has to be extended until it reaches it.

Cross-slope of formation

5 % (Exception: 2.5 % for bituminous blanketing layer

Ballast lowering below top of sleeper

Ballast bed lowering $v_{max} \geq 200$ km/h 4 − 5 cm

Fig. 2-56: Data on the ballast bed cross-section for plain line track in the ÖBB network [4]

In the same way as the thickness of the ballast bed, the dimension of the ballast shoulder also results from half the ballast width specified in the guideline and is between 45 and 50 cm depending on the sleeper length.

In Switzerland, the thickness of the ballast bed depends on the track loading group and the prevailing speed. [7, AB 25(3.2)] The distance between the bottom of the sleeper and the transition level between ballast bed and substructure below the lower lying rail is the ballast bed thickness. [7, AB 25(3.2)] The ballast shoulder in Switzerland is formed in a different way from Austria and Germany. The ballast profile at the shoulder is raised on purpose to increase the lateral resistance of the track. This increases the volume of the ballast in the shoulder and thus the ballast shoulder resistance, i.e. the lateral resistance to displacement. The distance to the maximum height at the ballast shoulder is measured horizontally from the track axis.

In the UK ballast depth is a function of track category (see Fig. 2-14) and varies from 300 mm for track in category 1A (240 kph) to 200 mm in category 6 (64 kph). Also, like Switzerland, a ballast shoulder is specified 125 mm above the sleeper end and with a width dependent on track radius and speed. [84]

Thickness of ballast bed

N1, E1	35 cm at > 160 km/h
	30 cm at ≤ 160 km/h
N2, E2	30 cm
N3, E3	30 cm
N4	25 cm
E4	20 cm

Horizontally measured distance to track centreline

Standard ballast bed cross-section	145 cm
Increased ballast bed cross-section	165 cm

Height of shoulder to top of sleeper

Standard ballast bed cross-section	10 cm
Increased ballast bed cross-section	15 cm

Ballast shoulder

2:3

Cross-slope of formation

5 % (Exception: 3 % for stabilised or bituminous barrier layers)

Ballast lowering below sleeper top

5 cm

Fig. 2-57: Data on the ballast bed cross-section for line track in the SBB network [14]

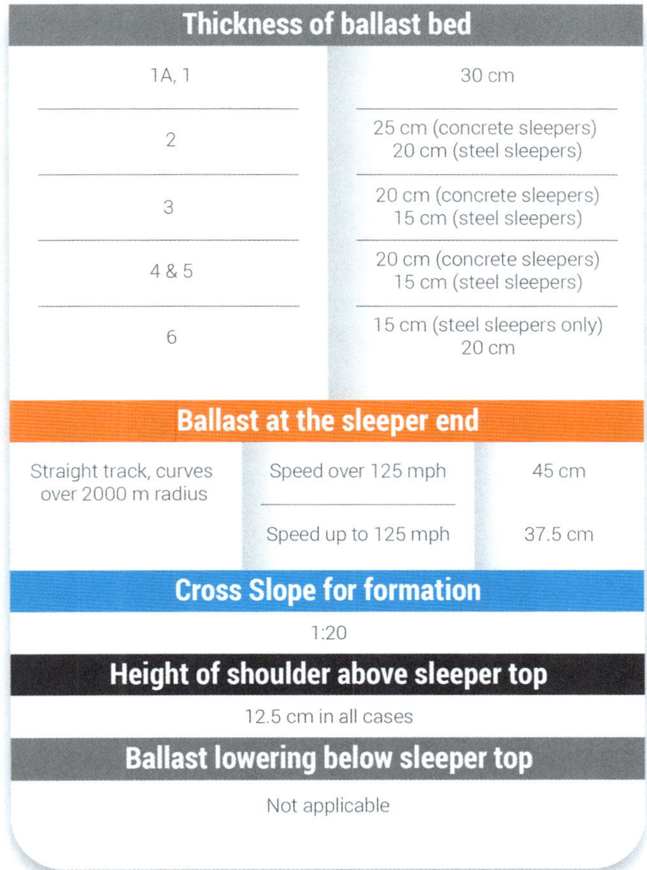

Fig. 2-58: Data on the ballast bed cross-section for plain line track in the UK network [85]

2.6.6   Substructure

The substructure or the subsoil is the final component in the vertical hierarchy of the track system for load transfer. Its importance can be compared to that of the foundations of a building. The role of this strata in the design of railways is frequently underestimated, apart from the knowledge that a soft subsoil with low bearing capacity can be disastrous. The terms 'substructure' and 'subsoil' are occasionally used interchangeably or confused with each other. Even though the distinction between track and substructure is not uniformly applied in the four countries, a distinction can be made between substructure and subsoil as follows:

– Substructure refers to all artificial (earth) structures which are designed to enable the routing of the railway line. In Switzerland, this also includes those artificial layers which sustainably improve a subsoil of low load capacity or deliberately change its characteristics. The Swiss Implementing Provisions of the Railway Regulations state:

"The substructure must have a good bearing capacity and low deformation and shall be dimensioned such that it can absorb the force transferred by the ballast bed without being damaged over its economic life (for new build 100 years as a rule; for maintenance projects [...] at least in accordance with the adjoining sections) and distribute this force at an acceptable magnitude to the subsoil." [7, AB 25(1.6)].
– In the UK the sub-structure is called the formation. Network Rail's Track Construction Standard states: "The design of the formation should take account of the subgrade material and the expected traffic loading. The formation should be stable and provide adequate support to the ballast layer to enable the required standard of track geometry quality to be maintained. Formation treatment should be designed to prevent ballast contamination by the migration of subgrade material and to direct water to the track drainage system. Where non-ballasted track designs are adopted the formation design should be carefully designed as subsequent adjustment of the slab after settlement can be difficult and costly. The design of any non-ballasted track should take the interface with existing assets and the provision of adequate transitions into account to minimise changes in vertical deflection under loading." [86]
– The subsoil comprises the in-situ soil and the bedrock. However, the top layer of the subsoil may well have been changed by construction measures. The subsoil, together with the substructure, forms the foundation of the track. [87, p. 30]

When planning railway lines, care is taken to adjust the route to the terrain as best as possible. Existing railway lines are, therefore, largely placed (up to 97% [87, p. 15]) on earthworks. During construction, attempts will be made to arrange earthworks (embankments, cuttings and side cuttings) in such a way that the soil masses are equalised as well as possible (see Fig. 2-59). If the gradient of the route layout is below the level of the natural land elevation, it is necessary to place the route into a cutting. The soil is excavated to the required level and side slope elevation for this purpose. Embankments frequently use, for example, the material excavated from cuttings to achieve the required height of the subgrade. Today it is important to check before use whether the excavated material has the required consistency and load bearing capacity. Recent embankment failures in the UK have shown that this was not always the case when railways were built in the 19th century!

In road and railway construction, gradient refers to the elevation of a route.

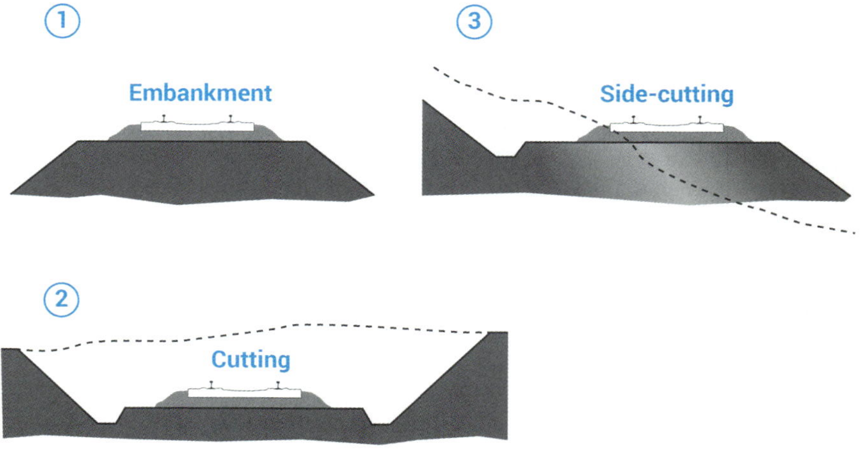

Fig. 2-59: Embankment, cutting and side cutting to permit the adaptation of the railway line to the terrain

A combination of both measures is used for cuttings built into the side of a hill to create the formation width required for the track. Earthworks are limited in their capabilities. Embankments have an enormous space requirement at their foot due to the safe angle of slope required as their height increases. For new lines, therefore, it is very difficult to overcome differences in height of several metres using embankments and civil engineering structures, such as viaducts, are chosen instead.

Today, for new lines, engineering structures such as tunnels and bridges are increasingly used (up to 50 % [87, p. 15]) to implement the planned route, particularly as the design speeds require larger curve radii, which makes it difficult for new lines to fit the terrain precisely.

To operate railway lines safely and sustainably both sufficient load capacity (fracture resistance) and suitability for use (deformation resistance) are required of the substructure, and the subsoil. As railways pass through large swaths of the landscape and frequently meet different natural conditions, getting the foundation right can have a major impact on ballasted track. Everything can be found, from hard rock to various types of unconsolidated material. The differences between very hard gravel of good bearing capacity and water retentive clays sensitive to deformation without additionally engineered strengthening measures are disastrous. In addition to such strengthening measures, various track bed structure systems or protective layers are frequently used to improve the bearing capacity of the substructure in order to keep deformation as low as possible and to ensure the load capacity of the railway line. These have a number of very different tasks, e.g.: [87, p. 295]

– load distribution
– frost protection
– separation and filter tasks
– sealing
– removal of rainwater

While bituminous layers (blanketing layer with binder) are used very often on new lines, a sand/gravel mix (blanketing layer without binder) is preferred for existing lines. Sand/gravel mixes can be installed by machine under normal operation and therefore using such machines has an advantage with regard to the construction time required.

These huge on track machines can install the required different layers (see Fig. 2-60) directly below the ballast bed. Similar to the ballast, they also have a defined grain structure, although it contains much smaller grains than the ballast in the track bed.

Fig. 2-60: Track cross-section with a protective sand/gravel layer

2.7 Switches and crossings (turnouts)

The great advantages of vehicle guidance along fixed rails also involves a few disadvantages. Railway vehicles are bound to their track, which makes the system inflexible. Thus, simple overtaking as we know it from driving a car is not possible. If two trains travelling at different speeds meet on a plain line, the faster train must either adjust its speed or find a way to change over to an adjacent track to pass the slower train.

2.7.1 Basic principles

To be able to change between tracks without interrupting travel, special structures in the track are required - the turnouts. The provision of turnouts creates flexibility and increases the capacity of the infrastructure. Using these, trains can travel on different tracks and change between them (Fig. 2-61).

Together with the track, turnouts form the basis for an efficient network infrastructure and have different purposes: [88, p. 13]

– formation and splitting up of trains
– utilisation of all tracks in stations
– crossing and overtaking of trains
– changing onto different parts of the track
– combining different tracks

Fig. 2-61: Groups of turnouts on ballasted tracks in Germany [89]

The first switches were used as early as the end of the 18th century in mine railways in England. [36, p. 239]

Designs published by John Curr, in Sheffield, England, as early as 1797 showed a "L" shaped switch plate with moveable tongue and different types of crossings to move colliery coal tubs from one track to another. [90, p.29] These form the basis for today's modern construction of turnouts, and their fundamental manner of operation has not changed, even after 220 years.

The term "turnout" is used as a generic name for a track unit that includes a switch and a crossing. There are many different types of switches and crossings; some of the most common shown at Fig. 2-62.

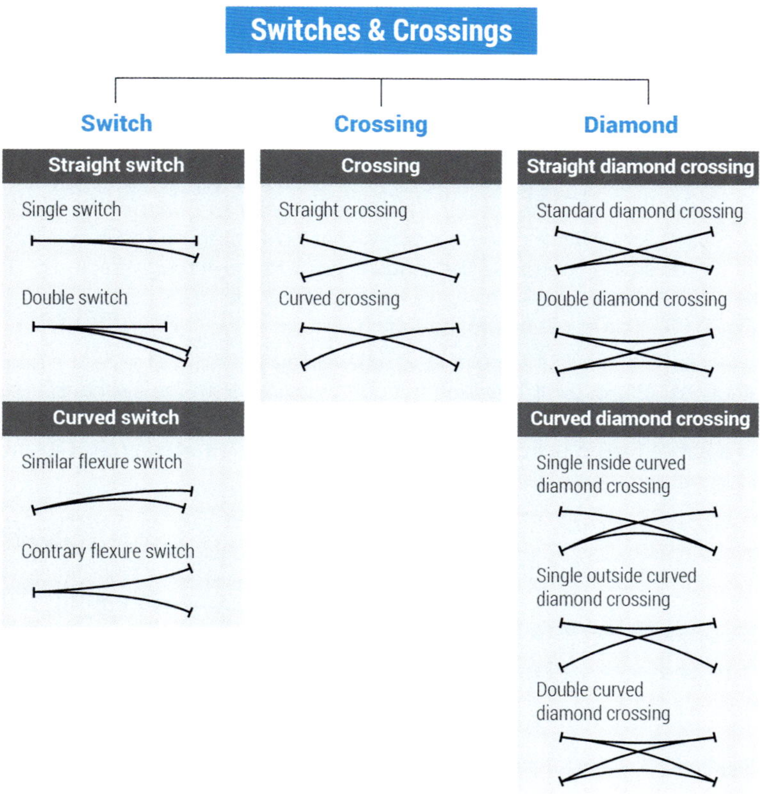

Fig. 2-62: Generic term "turnout" – classification of switches, crossings and diamond crossings

Crossings make it possible to travel across two tracks at the same level, but not to change from one to the other.

Diamond crossings combine crossings and switches. They, therefore, permit crossing and, depending on design, also the change of track with little space being required.

The special design features of turnouts give them special significance, both technically and financially. Turnouts form a disturbance (discontinuity) in a railway track. New developments aim to reduce the effect of this irregularity. Turnouts also lead to an increase in the forces acting on track due to the necessary breaks in rail surface continuity and thus to an increased maintenance requirement (more than 25 % of the total maintenance costs of a line [36, p. 240]). This, together with the susceptibility of turnouts to technical problems, contrasts with their operational necessity. Simply for the money spent on installing one simple turnout (EW 500), it would be possible to lay half a kilometre of plain track. [91, p. 29]

Even though the basic operation of switches in turnouts has not changed over the years, turnouts have been developed with differences in the four countries with regard to their design features and geometry. In Germany alone, there are more than 200 basic turnout types taking account of different rail types. [36, p. 243] This book describes only the technical principles based on a simple turnout and aims to create a general understanding of their basic manner of operation.

2.7.2 Terminology

A simple turnout enables a vehicle to be diverted from the through track to the diverging track and vice versa. Turnouts can be traversed in both directions. Despite this, a distinction is made between the switch toe and the switch heel together with the handing to provide uniform terminology. Switch toe (switch entry) describes that point of the turnout at which the track centreline of the diverging line forms the tangent point to the through track, i.e. the point at which the arc of the circle of the diverging track touches the track centreline of the through track (see Fig. 2-63). This is just before the switch rails.

The switch heel (switch exit) is that point at which welding or fishing of all rails is again possible without restrictions. [36, p. 249] Travelling through a turnout from the switch toe to the switch heel is described as "negotiating a set of facing points" and in the opposite direction as "negotiating a set of trailing points". If the divergence observed at the toe is to the right, the switch and turnout is right-handed and vice versa.

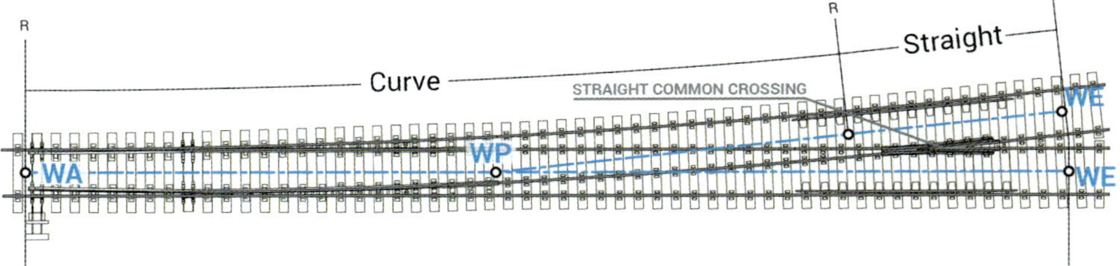

Fig. 2-63: Diagram of a single turnout

Similar to the curve of a plain line, the radius of the diverging track defines the permissible speed with which a train can travel through the turnout, or more accurately, onto the diverging track (see chapter 3).

Furthermore, the crossing angle characterises the special form of the turnout (see Fig. 2-66). It defines the angle or slope of the diverging track to the through track. Mathematically, it is formed from the tangent of the angle between the two track centrelines. This results in a crossing angle of 1 in 9 of the turnouts shown in Fig. 2-63.

When selecting suitable parameters, important cost considerations have to be taken into account, in addition to the permissible speed of the diverging track. Even if a large radius of the diverging line has advantages due to the higher permissible speed, it will entail high costs, not least due to the large space requirement. Therefore, besides operational considerations, costs must also be borne in mind for the correct choice of parameters.

2.7.3 Main components

A simple turnout is made up of different components. One distinguishes between the turnout bearers, the rail fasteners, the driving and locking mechanism and the rails.

The components of the turnout rails are divided again into the following three main groups (see Fig. 2-64):

– switch tongues or blades
– stock rails
– closure rail section
– crossing with check rails

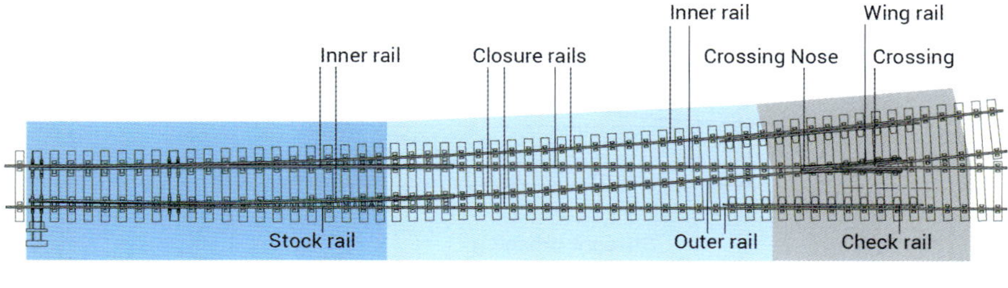

Fig. 2-64: Main turnout components

Fig. 2-65: Switch operating unit integrated into the train signalling system [89]

The switch tongue guides the vehicle either onto the through track or the diverging track (see Fig. 2-66). The switch rails, in contrast to the stock rails, are moveable and provide the guidance required. Due to their special shape, they sit closely against the stock rails and are moved to the required position during the switch operation over slide chairs. Various types of systems (hydraulic or mechanical) are used for the switching operation. The switching system both moves the tongues and secures them in their final position. Longer turnouts may have more than one linked switching system or drive mechanism. The locking systems required for this ensure that the switch rail sits tightly against the stock rail and, on the opposite side, that the switch rail is held away or open from the stock rail. Electronic detection systems are built into the locking mechanism such that when it is indicted that the final position has been established, the switch is released for travel. If the switch rail does not sit tightly against the stock rail, without a detection system, it could happen that one axle drives onto the diverging track while the others follow the main track. A derailment could not be prevented in this case.

The closure rails, which are curved, consist of standard rails which guide the vehicle to the crossing of the turnout and there transfer it to the wing rails. The crossing thus consists of a crossing nose and the two wing rails. To transfer from the through track to the diverging track, the through track must be gapped to permit the vehicle to cross. Therefore, the rail at the side of the crossing is gapped, creating a very short unguided point.

Due to the gap in the crossing, as the wheelset loses its guidance on one rail and could possibly be guided incorrectly. To prevent this, check rails are installed immediately opposite at these points. These make it impossible for the wheel to travel onto the wrong rail at the crossing. It should be noted that these are not designed to accept vertical loads, merely to steer the back of the wheel flange as it passes by.

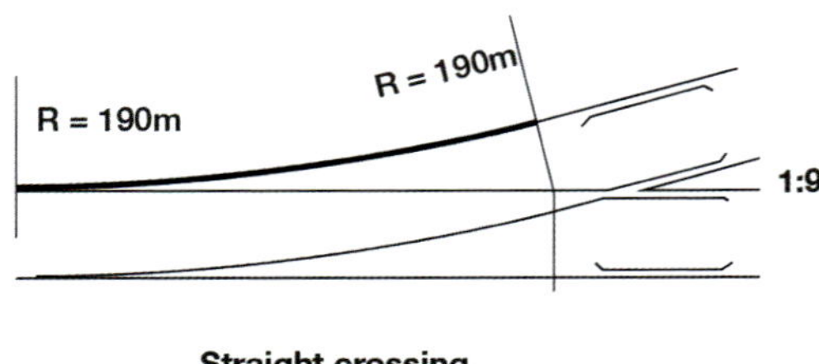

Fig. 2-66: Running edge diagram for a single turnout with straight crossing

A running edge diagram is the "top view of a turnout with only the running edges of stock rail and check rail shown, omitting other details". [88, p. 13]

There is one significant difference between the design of crossovers (two back to back turnouts) and major junctions between the UK and Europe as a consequence of the narrower spacing of adjacent tracks. This concerns the concrete or timber bearers in the area of the crossings. In the UK a simple two track crossover will have consecutive long bearers in the centre section between the two opposite crossings. This has a significant influence on the installation process and in the maintenance of track layouts when it comes to tamping.

Fig. 2.67: The long timbers in a UK crossover with track centres at 3405 mm. [92]

Summary

This chapter has given the reader a very brief insight into the history and development of railway track. We have looked at track as an engineering system and seen how four major European railways apply standards to its construction. We hope this will be useful as we develop and compare an understanding of the techniques used to design and maintain modern track in the following chapters.

3 Track alignment design

Fabian Hansmann and Richard Spoors

3.1 Key issues

This chapter describes the main features of longitudinal track alignment design and thus forms the basis for the presentation of the target track position. The different terminologies in the DACH countries and the UK are compared. Where possible, the terminology of European standards is used to ensure unambiguous and consistent terminology.

– Which basic alignment elements are used by railways?
– What key role does cant play and how can its physical effect be explained?
– What effects do different types of transition curve have on the running behaviour?
– What are the major physical differences between vertical and horizontal alignment elements?
– Which fundamental specifications limit the alignment elements in their geometric design?

3.2 Physical principles

The design of railway lines is based on fundamental physical relationships and is subject to the physical constraints of the track system. An understanding of these basic concepts creates an important added value for the delivery of maintenance work.

Trace refers to the railway track and its spatial location in the landscape. The trace thus represents the three-dimensional position of the track. To find the optimum trace for new railway projects or existing lines, a variety of criteria, such as required travelling speed, timetable, topography and the associated costs, have to be taken into account. Operational meanings of the term 'trace' are not covered in this chapter.

For straight line sections, the forces (G) that act upon the track are mostly vertical. Horizontal forces arise only to a comparatively small degree. Exceptions to this are lateral forces due to vehicle movement or special external influences such as wind. However, this changes when a train enters a curve. When travelling through a curve, the lateral forces increase depending on speed, vehicle behaviour and radius.

This chapter is concerned with the production of a smooth ride quality for passengers and the delivery of undamaged goods rather than the forces acting upon the track. If these criteria are not met, they can lead to derailment or overturning of the vehicle.

When a train moves continuously over the track, it attempts to maintain its current state of movement. Therefore, if one wants to accelerate or decelerate a train or change its direction of travel, a corresponding force must act on it. Any passenger can notice this phenomenon. When the train accelerates, one's own body tends to remain stationary relative to the train. In contrast to this, the body tries to maintain the original motion when the train is braking. In both instances, one has the feeling that the body wants to move in the opposite direction to the train. In mechanics, this principle is referred to as Newton's law of inertia and applies to every change of direction of a vehicle.

The principle of inertia can be illustrated by the example of a hammer thrower (Fig. 3-1). Due to the force of the hammer thrower and his rotary movement, the hammer starts to move in a circle round the thrower and is kept in its path. When the hammer is released, it continues to move on a straight line in accordance with the law of inertia.

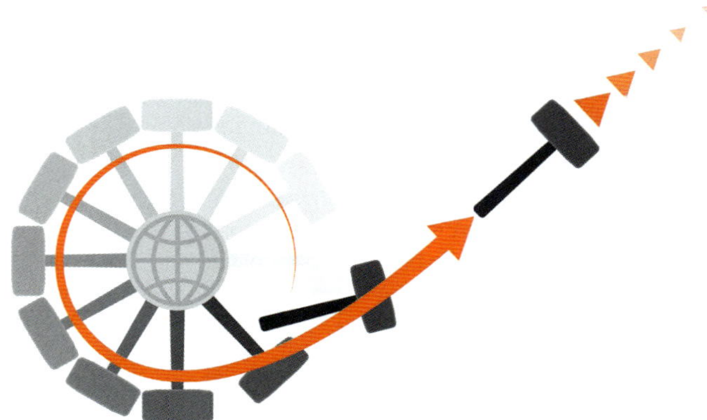

Fig. 3-1: Illustration of the law of inertia by the example of a hammer thrower

Due to the change in direction when travelling through a curve, a force F directed to the outside of the curve acts on the passenger or the load (see Fig. 3-2). This force is generally referred to as centrifugal force and represents a pseudo-force which only acts on objects in the vehicle.

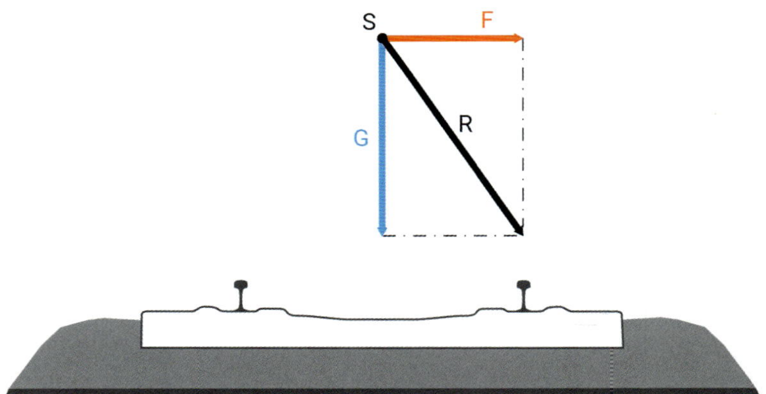

Fig. 3-2: Forces F and G at the centre of gravity of the vehicle body and the resultant force R

Force F in [N] results generally from mass m in [kg] multiplied by acceleration a in [m/s^2]:

$$F = m \cdot a \qquad (3\text{-}1)$$

Therefore, a force

$$F = m \cdot \omega^2 \cdot R \qquad (3\text{-}2)$$

acts on objects in the vehicle with ω in [m/s] as angular velocity:

$$\omega = \frac{v}{R} \qquad (3\text{-}3)$$

$$F = m \cdot \frac{v^2}{R} \qquad (3\text{-}4)$$

The use of equations in railway practice has one peculiarity. While normally numerical values with the same unit are used (e.g. length stated in metres throughout), in railway practice quantities with different dimensions can be combined. To avoid calculation errors, it is then necessary to take account of correction factors or conversion factors. The equation for centrifugal force combines the radius of the curve with the speed of the vehicle. Usually, the radius of curvature is stated in metres and the speed of a train in kilometres per hour. To convert from km/h to m/s, the following factor is used: $\frac{1}{3.6}$ ($\frac{1000}{60 \cdot 60}$).
Therefore, the equation changes $\frac{1}{(3.6^2)}$ to:

$$F = m \cdot \frac{v^2}{12.96 \cdot R} \qquad (3\text{-}5)$$

This peculiarity makes it necessary to be absolutely clear before carrying out any calculation about which units individual variables will use in the calculation.

To find a suitable alignment, the resulting forces are not as important as the acting acceleration. From the force calculated, the following is obtained from the known relationship between force and acceleration from (3-1):

$$a_Q = \frac{v^2}{12.96 \cdot R} \qquad (3\text{-}6)$$

v in [km/h]: speed

R in [m]:   radius

a_Q in [m/s]: lateral acceleration

The value for lateral acceleration is limited for reasons of ride quality, dynamic safety and maintenance expenditure. While higher values can be tolerated with a view to dynamic safety, the ride quality can be severely impacted if passengers are subjected to high acceleration or if this changes rapidly. In the DACH countries, as a rule, the normal value is 0.98 m/s^2 (Δd = 150 mm). [1] – [3] This limit value has to be reduced accordingly to take account of the quality of the track, the vehicle stock and the requirements on ride quality as well as maintenance. In isolated cases, e.g. at turnouts, this limit can certainly be exceeded, but this should be locally restricted.

In the UK, the value of acceptable lateral acceleration in the design of curved track is not explicitly stated in Network Rail's Track Design Handbook, however, the value of 0.98 m/s² has been allowed for in the factor for the calculation of equilibrium cant, which is 11.82. Equilibrium cant is the value of cant for a given radius where the outward acceleration at a stated speed is equal to the inward acceleration caused by the cant. [4]

If a line with a target speed of 160 km/h is planned, taking into consideration a lateral acceleration of 0.98 m/s², this results, in accordance with (3-6), in a minimum permissible curve radius as follows:

$$R = \frac{v^2}{12.96 \cdot a_Q} \tag{3-7}$$

$$R = \frac{160^2}{12.96 \cdot 0.98} \approx 2000 \text{ m} \tag{3-8}$$

Modern railway lines certainly use curves with large radii, but their target speeds are frequently twice as high. A simple trick helps to reduce the radius required. By changing the elevation of the opposite rail, the reference plane tilts towards the inside of the curve. Thus, the rail on the outside of the curve is higher than the one on the inside. This measure is referred to as cant or superelevation (Fig. 3-3) and stated as the height difference d in [mm] between the two rails at the running edges S in [mm].

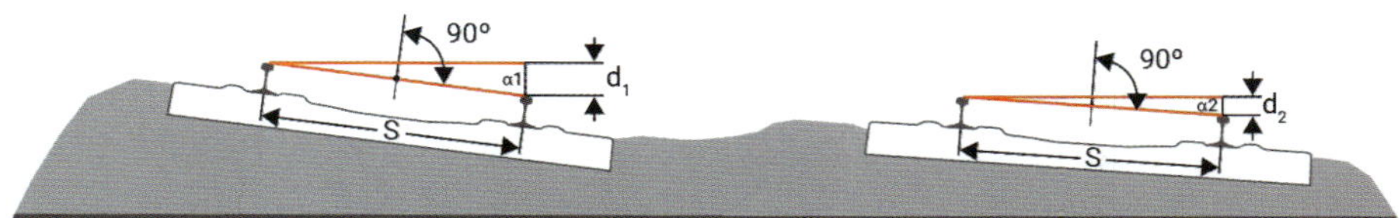

Fig. 3-3: Differences in the cant of two tracks

Thus, on entering the curve, the vehicle body tilts in the direction of the centre of the curve (inside of the curve). This is a phenomenon which is also used for banked corners in motor racing to increase the speed in curves (see Fig. 3-4).

Fig. 3-4: In motor racing, banked corners permit driving at high speeds [5]

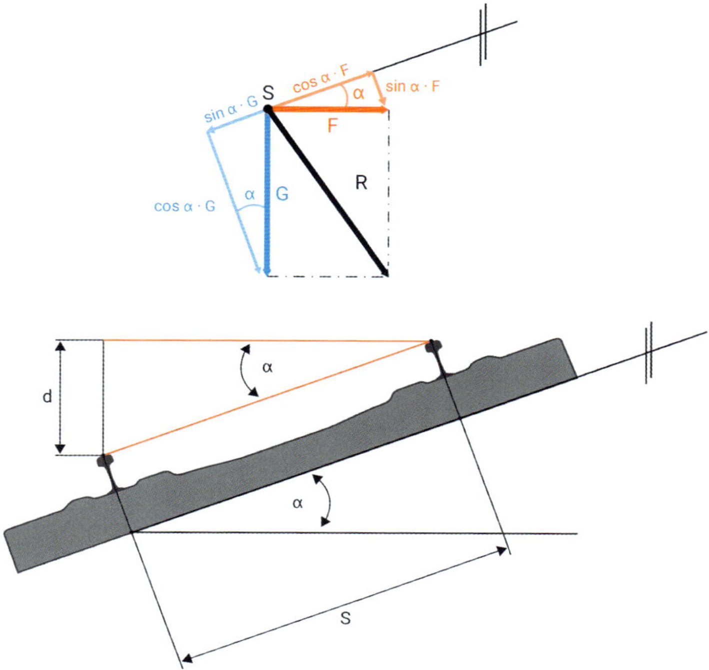

Fig. 3-5: Diagram of forces due to the inclination of the reference plane

Due to this small adjustment, the equilibrium of forces shown in Fig. 3-2 shifts (Fig. 3-5). Although the acting forces remain the same, they are now divided differently. The sum of the forces acting in the direction of the outer rail is as follows:

$$F_B = \cos(\alpha) \cdot F - \sin(\alpha) \cdot G \qquad (3\text{-}9)$$

The forces are reduced due to the two trigonometric functions. Furthermore, part of weight force G now counteracts the centrifugal force. Angle α has very small values ($< 6°$) in the case of ballasted track. Thus, equation (3-9) can be simplified.

$$\cos(\alpha) = 1 \qquad (3\text{-}10)$$

$$\sin(\alpha) = \frac{d}{S} \qquad (3\text{-}11)$$

$$F_B = 1 \cdot F - \frac{d}{S} \cdot G \qquad (3\text{-}12)$$

$$a_Q \cdot m = 1 \cdot m \cdot \frac{v^2}{R} - \frac{d}{S} \cdot m \cdot g \qquad (3\text{-}13)$$

$$a_Q = \frac{v^2}{R} - \frac{d}{S} \cdot g \qquad (3\text{-}14)$$

Taking account of gravitational acceleration at 9.81 m/s² and the relevant units, the following is obtained:

v in [km/h] R in [m] a_Q in [m/s] d in [mm]

$$a_Q = \frac{v^2}{12.96 \cdot R} - 9.81 \cdot \frac{d}{S} \tag{3-15}$$

Assuming a distance between the running edges of S [6, p. 3] of 1500 mm (gauge = 1435 mm), the equation can be simplified as follows while still being sufficiently accurate:

$$a_Q = \frac{v^2}{13 \cdot R} - \frac{d}{153} \tag{3-16}$$

By transforming the equation, the cant at which the acting acceleration a_Q is completely compensated, i.e. takes the value of zero ($a_Q = 0$), can now be easily calculated.

$$d = \frac{11.8 \cdot v^2}{R} \tag{3-17}$$

This cant is described as the theoretical or equilibrium cant. However, the reduction of the acting forces applies only to trains with the selected speed in a curve with radius R. As trains travel over the same section of a mixed traffic line with different speeds, a suitable compromise has to be found between slow goods trains and fast long-distance passenger trains. Even on lines on which all trains travel at the same speed, it is always possible that operational changes occur, which have to be accommodated (e.g. unplanned train stops due to signalling).

In addition to the normal acceleration of a train, any sudden change is of major importance for the ride quality. A change in acceleration can lead to nausea for passengers, similar to a rollercoaster, and is regarded as annoying. This change is called a jolt and is felt as a shock inside the vehicle. However, in addition to adjusting the alignment to the topography as well as possible, it is also possible to design the required changes in direction to be smooth, while maintaining the ride quality, by arranging various design elements. Differences in the acting lateral acceleration can thus be avoided or mitigated as much as possible by means of various design elements.

3.3 Components of a railway route

Before the actual construction starts on a new railway line, different routing options are checked with regard to their implementation. Different design elements are available for different routes depending on parameters such as projected speed, vehicle composition and the type of operation.

In addition to modern 3-D virtual representations, the classic plan view as a layout plan and contour map continues to be used today. While the layout plan shows the overview of the route (top view), the contour map shows the gradients and longitudinal profile.

The design elements are specified in accordance with EN 13803 as "segment of the track with either vertical direction, horizontal direction or cant" which obey "a unique mathematical description as function of longitudinal distance". [7, section 3.4]

As per EN 13803, design elements can be differentiated into elements of the layout, elements of the elevation and elements of cant, depending on the type of view. [7]

For the layout plan (Fig. 3-6), this gives the following elements:
- straight line
- curve
- clothoid type transition curve
- alternative transition curves

For the elevation:
- constant superelevation
- parabola curvature
- circular

As well as elements of cant:
- constant cant
- linear cant ramp or gradient
- non-linear cant ramp or gradient

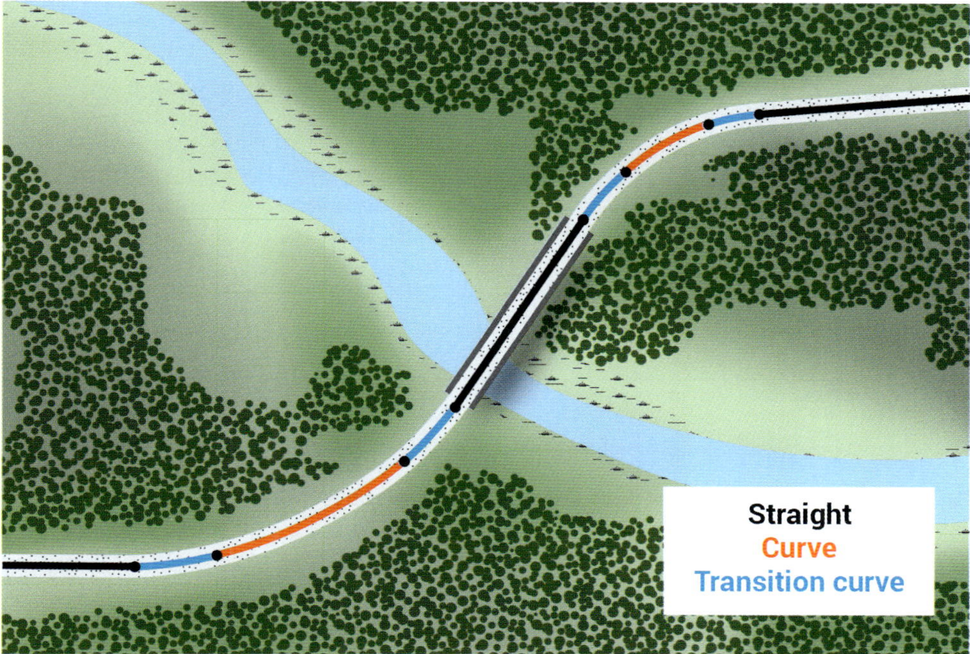

Fig. 3-6: Design elements of planned layout with increasing kilometrage from left to right

3.4 Limits and standard values

In general, design elements are limited not only by the existing space, but also by the values and limits in national standards. These are defined in the rules and standards of individual railway companies or in European standards and laws.

The difficulty in quoting standard values and limits is in the fact that the individual DACH countries and the UK differentiate them differently.

In Germany, RIL 800.0110 differentiates between the discretionary range and the approval range. Values in the German approval range (approved values and EBO limits) are to be avoided in principle and require special approval by head office of DB Netz AG (currently: Dynamics of Vehicle Movement/Line Layout Division). In principle, the standard values have to be observed for new construction. However, it is incumbent upon the person responsible for the region to utilise the discretionary range up to the discretionary limit. However, such utilisation must be justified with economic and technical arguments. [8]

If values are compared in this book, these values are the standard values in the case of Germany.

In Austria, a distinction is made between recommended and exceptional limits. The usual rule is to observe the recommended limit. Exemption permits have to be granted by the technical department in charge (Line Management and Installation Development Division, ÖBB Track Technology) for exceptional limits. Exceptional limits may be exceeded only if this does not result in any safety-critical restrictions. Supplementary to the other DACH countries, ÖBB standard 01.03 further differentiates between the maintenance and the new construction of TEN lines (lines of the trans-European Railway System) and other lines. This book only quotes the recommended limits for TEN lines of track class a. [9]

Germany	Austria	Switzerland	United Kingdom	EN 13803
Maximum and minimum value	**Recommended limit**	Planning limit	**Normal limit**	**Normal limit**
Standard value	Exceptional limit	**Normal limit**	Exceptional limit	Exceptional limit
Discretionary limit	EisbBBV limit	AB-EBV limit		
Approved value				
EBO limit				

Fig. 3-7: Overview of the types of design limits for track geometry [7] – [10] [11]

In Switzerland, a distinction is made between planning limits and normal limits. Limits are not defined solely for safety reasons but can also be justified due to maintenance or economic considerations. The planning limit is broader than the conservatively set normal limit. In principle, planning limits must be observed for new lines and are to be aimed at when modifying existing installations. Normal limits can be used, if they are required, without any special additional measures, although making use of the broader limits leads to higher maintenance costs. The limits of AB-EBV must be observed without fail. This book quotes the normal limits unless otherwise specified. [10]

In the UK, Network Rail, like ÖBB, makes a distinction between recommended and exceptional limits. The reason and implication for the use of any limit beyond those recommended up to the exceptional limits needs to be documented in accordance with the standard "Engineering Assurance Arrangements for Track Engineering Projects". [12]

EN 13803 refers to normal limits and exceptional limits: "These exceptional limits represent the least restricting limits applied by European railway companies and are intended exclusively for the application under special circumstances and may require a corresponding maintenance system." [7, section 3.3]. The standard advises in principle to avoid making use of the complete extent of limits, in particular exceptional limits. As for the limits of railway companies, the normal limit is used where limits from the standard are referred to in this book.

Figure 3-7 shows a comparison of the different national limits. The values which are used in this book (unless stated otherwise) are shown in bold.

A straight line as the shortest connection between two points is the simplest design element of the trace for a railway track. As the forces acting on the vehicle and the track are lowest here, the objective will be to lay the track as straight as possible. However, it is impossible to implement a railway track by only using straight lines. This would be either technically impossible or too expensive due to the construction activities required (e.g. bridges or tunnels).

Curves are used as necessary to change the direction. They make it possible to fit the trace closer to the existing landscape and to avoid obstacles such as hills or protected areas. Simple curves are differentiated into left-hand curves and right-hand curves. In a left-hand curve the direction of travel deviates towards the left with increasing kilometrage. The same applies to a right-hand curve, but in the opposite direction. If a left-hand curve is followed by a right-hand curve or vice versa, this is called a reverse curve. If two consecutive curves have the same direction, this is described as a compound curve (see Fig. 3-8), even if they are separated by a transition.

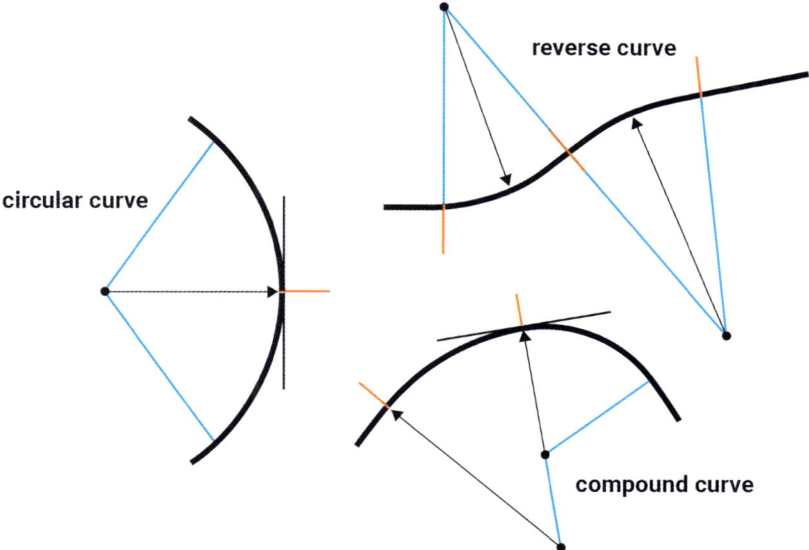

Fig. 3-8: Curves and their different combinations

In principle, it is possible to transfer from a straight line immediately into a curve. In urban areas this is sometimes done for very tight tramway curves which are taken at low speeds. The only thing to be observed is that the straight line forms the tangent to the curve at its contact point. Despite this, a jolt arises due to the sudden change in the radius of curvature, which has a negative effect on ride quality and must be avoided on main-line railways as a matter of principle. The radius of the curve (i.e. its curvature) is a function of the planned speed of trains and is quoted in all DACH countries and the UK as R in [m]. The current TSI Infrastructure lists the minimum radius for new lines as 150 m. [13, section 4.2.3.4]

Abbreviation of radius:

Germany: r   Austria: R   Switzerland: R   UK: R

The curvature k of a curve determines its deviation from the straight line and is defined as the inverse of radius R for curves. The curvature is usually shown below the contour map in a curvature diagram. This maps the curvature in relation to the length of the track, with the curvature of left-hand curves being plotted downwards (negative) and the curvature of right-hand curves being plotted upwards (positive) as a rule.

As radii are quoted in metres for railways and generally take a large value, the curvature value will be very small. However, to be able to make these values clearer in a curvature band, the calculated value is multiplied by 1000.

$$k = \frac{1000}{R} \tag{3-18}$$

R in [m]:   radius

k in [1/m]: curvature

There is a great advantage in determining the radius of curvature through the curve: Mathematically, straights have an infinite radius which makes it impossible to show this in a diagram. The curve does not have this disadvantage. It can be plotted easily over the trajectory of a line (Fig. 3-9), and the curvature diagram created in this way helps to identify the various design elements easily. Each curve is bounded by the starting point of the curve (BA) and the end point of the curve (BE) in relation to the line kilometrage. The main design points are always identified in the direction of increasing kilometrage.

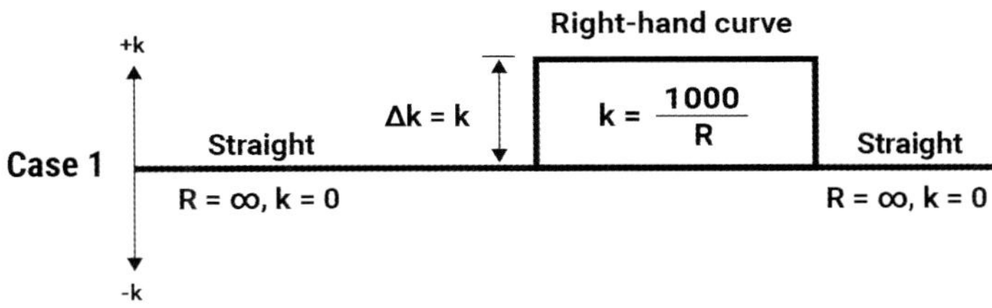

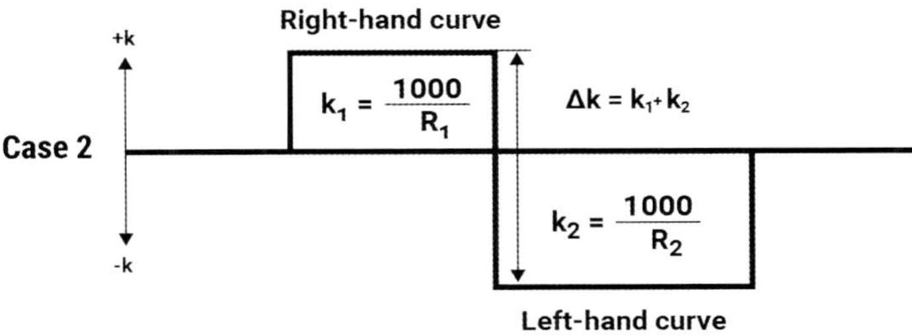

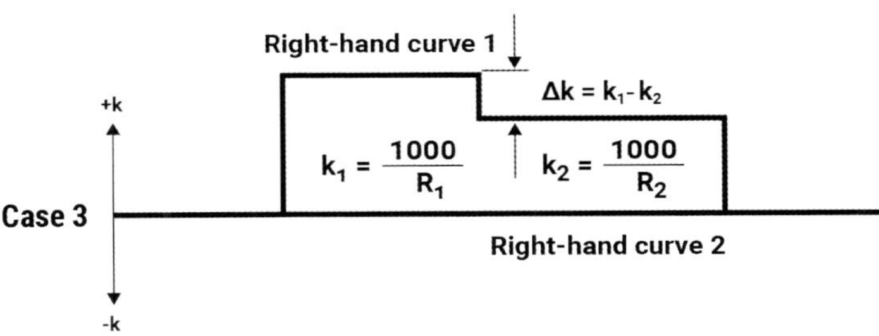

Fig. 3-9: Curvature diagram – Case 1: simple curve, Case 2: reverse curve, Case 3: compound curve

3.5 Transition curves and cant ramps

Transition curves are elements of track alignment used to smooth the movement of the vehicle into a curve and to minimise the jolt. They have the task of reducing the jolting motion between straight and curved track horizontally (transition curve) and vertically (cant ramp or gradient). While the transition curve adjusts the radius of the curvature from zero (value of the straight) to the value of the curve, the ramp or cant gradient has the task of steadily increasing the value of cant to that of the regular curve.

Special curves whose curvatures increase linearly with length are normally used for transition curves. [14, p. 25] Therefore, the curvature at the start equals zero and increases to the required radius at the end. Thus, the transition curves adapt geometrically as well as possible to the straight at the one end and the curve at the other end. In this way a transition between two curves is created. There are different curves which meet this requirement. For both railway and road construction it is the clothoid that is mainly used.

Clothoids have the special characteristic that their curvature increases proportionately with their length. The clothoid has the characteristic shape shown in Fig. 3-10. In nature, this shape is known from the snail shell. While one end approximates closely to a straight line, the other end permits variable adaptation to the radii required.

Fig. 3-10: Clothoid in nature [5]

The setting out dimensions of a clothoid can be determined accurately by means of a mathematical series expansion. Setting out dimensions refer to those points of the curve which are used to transfer the design of a curve to the construction site. It was difficult to determine this for clothoids without using a computer, so a sufficiently accurate approximation can be substituted. A cubic parabola is frequently used for this purpose even today:

$$y = \frac{x^3}{6 \cdot l \cdot R} \tag{3-19}$$

l in [m]:   length of parabola

R in [m]: radius of adjoining circular curve

This type of transition curve is used as the standard solution in the DACH countries and in the UK. Fig. 3-11 shows the arrangement of a clothoid as the transition curve. It can be seen that, when using a transition curve, the straight line – in contrast to the direct transition between straight line and curve – does not form the tangent to the circle. The curve is displaced from the straight line by offset f and distance a. It can be estimated using the parabola as follows:

$$f = \frac{l^2}{24 \cdot R} \tag{3-20}$$

$$a = \frac{l}{2} \tag{3-21}$$

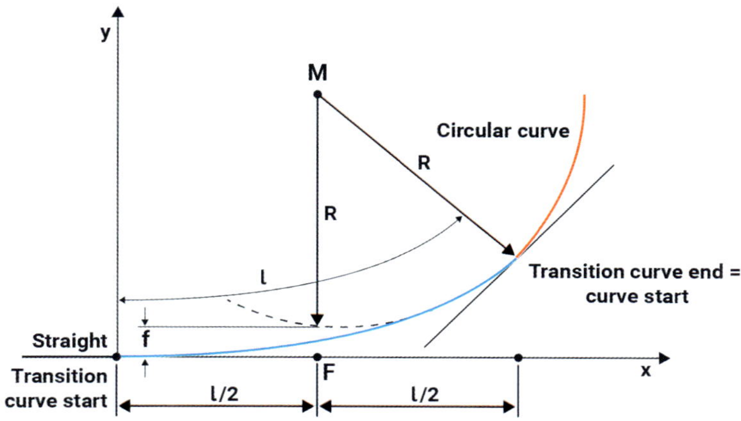

Fig. 3-11: Diagram of the plan view of a clothoid as a transition curve

Every transition curve is uniquely defined through the start of the transition curve and the end of the transition curve. The length of the transition curve (l) depends on its shape. As with the curve, the increasing curvature of the transition curve is plotted in the curvature diagram specific to the direction (left-hand = negative/downwards; right-hand = positive/upwards) (see Fig. 3-12).

3.5.1   Transition curves and linear cant ramps

The cant increases almost linearly to the required value in the same way as the curvature of the transition curve. This is termed a linear cant ramp or gradient. In general, care has to be taken that the cant ramp and the transition curve have the same length. Exceptions are noted in the rules and standards of the particular railways. The start and end of the cant ramp are shown in Fig. 3-12 by the bottom and top of the sloping lines.

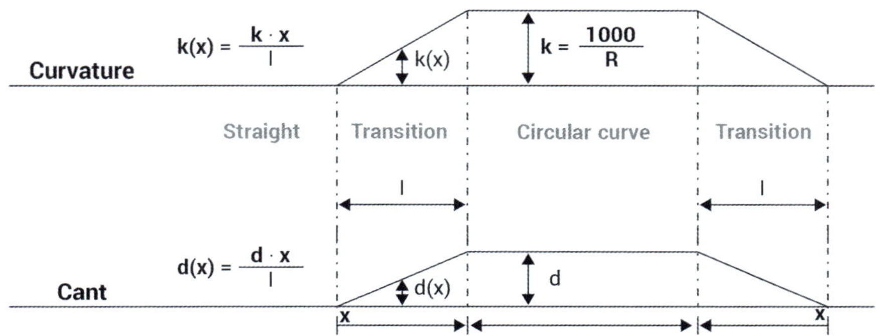

Fig. 3-12: Development of cant and curvature for a curve with linear cant ramp in Austria and Germany

Symbols for cant:

Germany: u in [mm]   Austria: d in [mm]   Switzerland: ü in [mm]   UK: E_a in [mm]

However, there is one significant difference in the design of cant d between the countries: In Germany, Austria and the UK, the cant is created by lifting the outer rail by the required amount (Fig. 3-13 (left)). In the case of a linear ramp, e.g. for a left-hand curve, the right-hand rail is raised continuously to the required cant value from the start of the cant ramp. The left-hand rail remains in its original position. The opposite applies to a right-hand curve. It is not common to lower only the inside rail in a curve to create the cant.

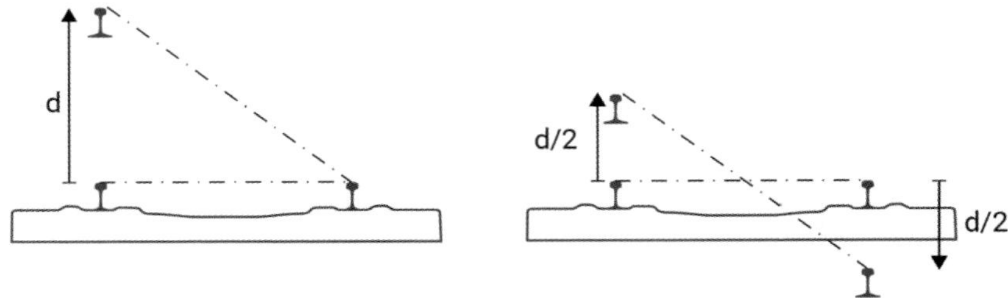

Fig. 3-13: Creation of cant by lifting the outer rail only (left) or moving both in opposite directions (right)

In Switzerland, both rails are rotated around the intersection between the track axis and running surface, independent of the direction of the curve. For this purpose, the outer rail is lifted by half the required amount, while the inner rail is lowered by the same amount (Fig. 3-13 (right)). This opposing change in the elevation of the rails is plotted accordingly in the cant diagram (see Fig. 3-14).

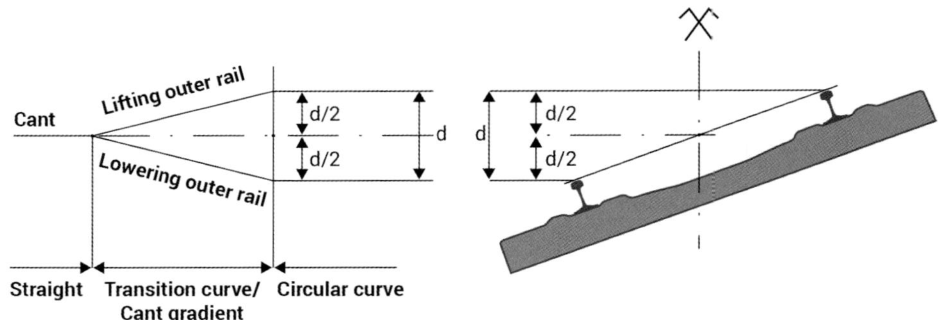

Fig. 3-14: Creation of cant in Switzerland by adjusting both rails [15]

This procedure has the advantage that the centre of gravity of the vehicle is kept approximately at the same level, because lifting the outer rail (as is common in Germany, Austria and the UK) shifts the centre of gravity upwards by d/2 when travelling over the cant ramp.

A further development of this method is the Wiener Bogen® (Viennese Curve). This was developed by ÖBB as a special centre of gravity layout. Here, the rails rotate around the centre of gravity of the vehicle (Fig. 3-15) which is meant to create smooth vehicle motion. However, by rotating the rails, their alignment in the layout plan also changes. Further details on this rare type of transition curve are given in the relevant ÖBB guideline (01.03 Linienführung [16]).

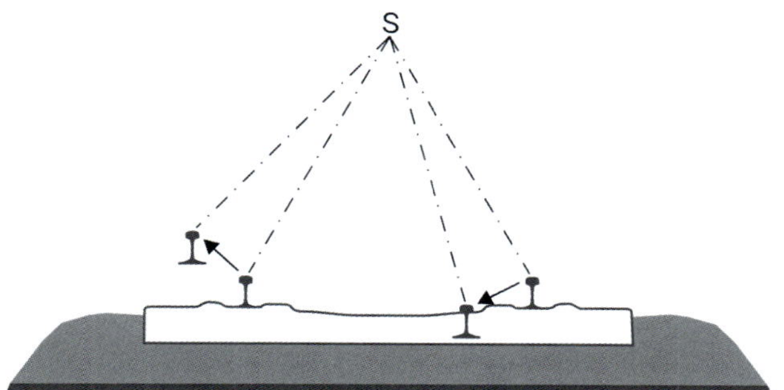

Fig. 3-15: Diagram of the principle of the "Viennese Curve" centre of gravity layout

A linear cant gradient is quoted either as a ratio x:1000 or in [‰].

Fig. 3-16: Calculation of the cant of a continuously increasing gradient of ratio 1:1000

In the UK the design of transition curves may be based on the rate of change of cant as well as cant gradient. Network Rail's Track Design Handbook states: "For permissible speeds up to 50 mph (80 kph) and enhanced permissible speeds up to 70 mph (120 kph), design of transitions will normally be determined based upon cant gradient rather than rate of change of cant. Both of these guideline speeds are based on maximum design values." [17]

In addition to the linear cant ramp, other cant ramps and transition curves have been designed in the DACH countries. They are meant to either increase the ride quality or to enable a more economical increase in speed due to a shorter transition length.

3.5.2 Transition curves and graduated cant ramps

Around 1937, Gerhard Schramm developed the graduated transition curve to make the transition between straight track and curve smoother. Originally, this transition was designed as early as 1870 by Friedrich Robert Helmert. [18, p. 621]

The graduated transition consists of a straight gradient with a vertical curve at both inflexion points with curves (Fig. 3-17) which touch in the centre of the transition. The graduated cant ramp deviates in its elevation by 19 mm from the straight variant. Different curves were used for the vertical curve, such as parabolas or sine waves; however, all led to a comparable result. Therefore, a quadratic parabola was adopted for the vertical curve due to the easy calculation, even though a sine curve was recommended for the best design. [6, p. 21]

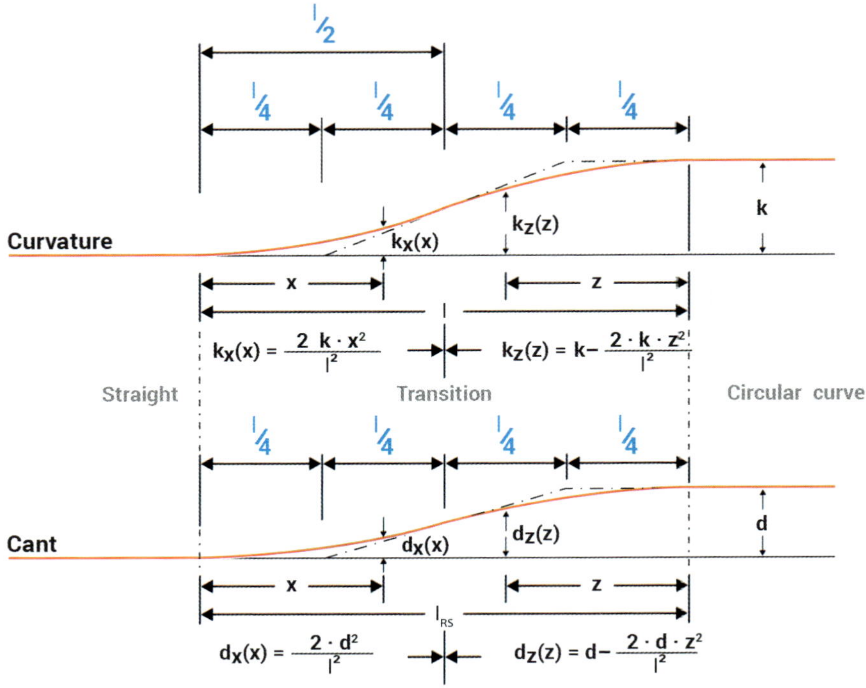

Fig. 3-17: Graduated transition curve based on Schramm

Referring to Fig. 3-17, the vertical curves of the graduated transition versines (kx and kz) continue into the curve and the adjoining straight. If an existing linear transition curve is to be replaced by a Schramm transition curve, one has to be aware that this requires considerably more longitudinal space if the cant and cant gradient are to remain the same.

The curvature diagram of the Schramm transition curve includes a graduated transition curvature and a graduated cant ramp, both of which are calculated using the same formulas. If the length of the transition curve or the cant ramp l is divided into four sections of equal length, this results in the characteristic diagram of the Schramm transition.

However, the calculation of individual cants and curvature is more difficult than for a linear transition curve. Due to the opposite direction of curvature of the vertical curves, these values must be calculated either from the start to centre x or from the end to centre z.

50 % of the offset of the Schramm transition results from the offset of the clothoid and is calculated as follows:

$$f = \frac{l^2}{48 \cdot R} \tag{3-22}$$

To avoid a sudden transition between adjoining alignment geometry and the cant ramp, the transition areas in Austria must be curved in a similar way to the graduated Schramm transition. [16] However, unlike the Schramm transition, this graduated ramp uses two circular curves. The portion of the ramp between the two vertical curves continues to be connected by a straight of the original gradient. The vertical curve aims to improve the running behaviour of trains but does not have any effect on other parameters, such as permissible speed or the cant itself. This measure applies only to the values of cant in the transition curve. The curvature of the transition curve is not affected, in contrast to the types of graduated transitions common in Germany.

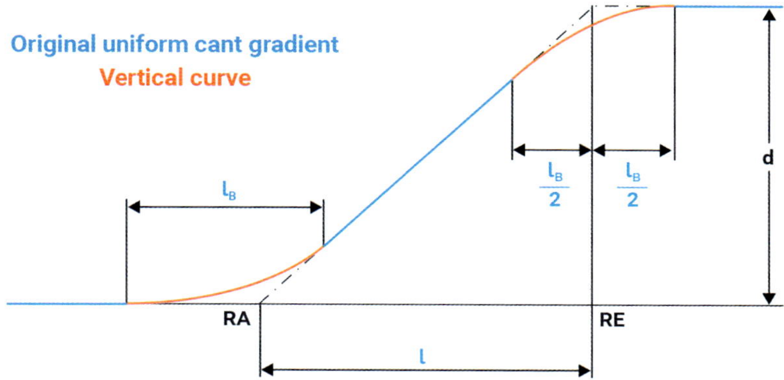

Fig. 3-18: Vertical curves at the inflexion points of the cant ramp in Austria [16]

The approximate dimension is specified in the guideline and depends on the cant gradient in the ramp, while the length of the vertical curve is calculated as follows:

$$l_B = 110.0495 \cdot \sqrt[3]{\frac{|\Delta d|}{l}} \text{ at least } l_B = 10 \text{ m} \tag{3-23}$$

In Switzerland, no vertical curve is provided for in the transitions.

A slightly different type of graduated transition curve was suggested for the first time by A. E. Bloss in 1936. [18, p. 622] In contrast to Schramm, Bloss did not use two quadratic parabolas, but one continuous cubic parabola for the vertical curve. When comparing the progression of the two graduated transitions (Fig. 3-17 and Fig. 3-19), the more favourable characteristics of the Bloss transition become obvious: It rises noticeably more in its first third so that the length of the transition curve does not have to be extended into the adjoining alignment geometry. In the centre, the polygon takes a constant progression to approximate to the desired value at the end. As the transition does not have to be split into two parts, cant and curvature can be calculated in single steps.

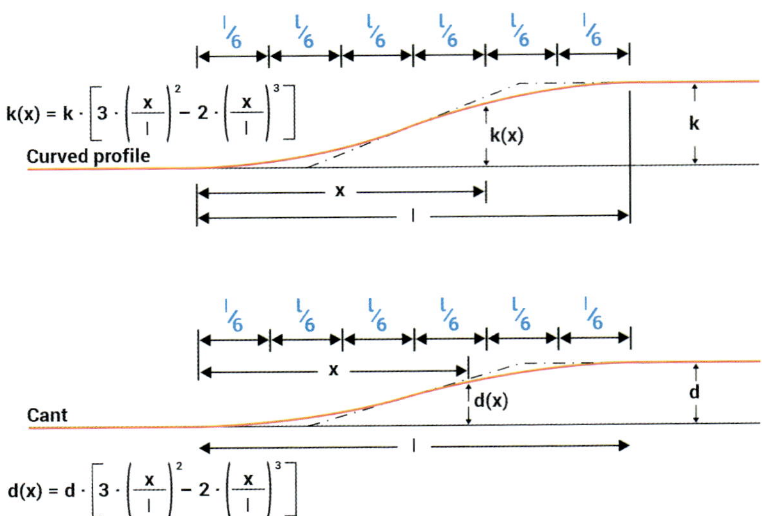

Fig. 3-19: Graduated transition curve based on Bloss

The offset of the Bloss transition is calculated as follows:

$$f = \frac{l^2}{40 \cdot R}$$

(3-24)

In Germany, existing Schramm transition curves are being replaced by Bloss transition curves during maintenance. [8] Even if dynamic measurements cannot show a detectable difference between linear and graduated ramps [18, p. 623], graduated Bloss transitions have important advantages which result from the cant ramp profile and the resulting length of the transition curve.

In contrast to the linear cant ramp, the gradient of a graduated cant ramp is variable. This results from the tangent to the centre of the ramp R_m. The cant gradient of a Schramm ramp is calculated from the tangent to the parabola in the centre of the ramp. In this case it is obviously not important to which of the two parabolas the tangent is applied.

Linear cant gradient	Schramm cant gradient	Bloss cant gradient
$\dfrac{1}{m} = \dfrac{d}{l_R}$	$\dfrac{1}{m} = \dfrac{2 \cdot d}{l_R}$	$\dfrac{1}{m} = \dfrac{3 \cdot d}{2 \cdot l_R}$
$m = \dfrac{l_R}{d}$	$m = \dfrac{l_R}{2 \cdot d}$	$m = \dfrac{2 \cdot l_R}{3 \cdot d}$

Fig. 3-20: Calculation of different cant gradients

d in [mm]: cant

l_R in [m]:   length of cant ramp

m in [-]:   gradient

The advantage of the Bloss transition is that its length can be shorter than the clothoid. The standard values for the length of the cant ramp result from the ideal vehicle movement as a function of speed. Therefore, in Germany, m = 10 · v [6, p. 14] applies to linear ramps and m = 4 · v [6, p. 21] to graduated ramps.

The standard lengths for cant ramps for the discretionary limit specified in the guidelines of Deutsche Bahn are calculated from this as follows:

Linear cant gradient	Schramm cant gradient	Bloss cant gradient
$l_R = \dfrac{8 \cdot v \cdot d}{1000}$	$l_R = \dfrac{8 \cdot v \cdot d}{1000}$	$l_R = \dfrac{6 \cdot v \cdot d}{1000}$

Fig. 3-21: Comparison of lengths of cant ramps of Deutsche Bahn in accordance with [8]

Symbols for cant gradient:

Germany: 1:m   Austria: D‴   Switzerland: N   UK: there is no specific symbol

In addition to designing for smooth vehicle movement, the permissible cant gradients (Fig. 3-22) have to be taken into account when designing the transition curves. If the cant increases too strongly, the difference between two consecutive axles or bogies can no longer be absorbed by the vehicle, and this leads to derailment.

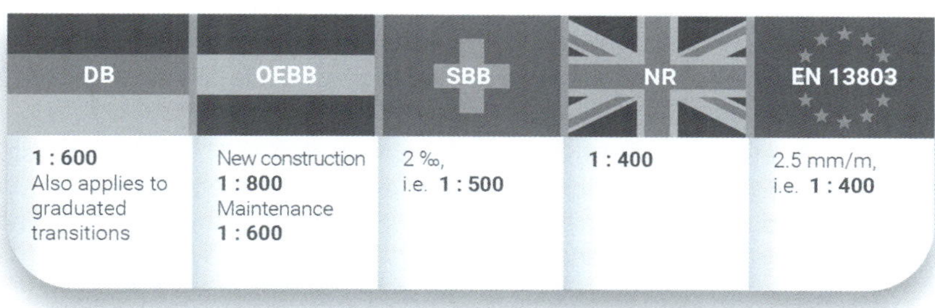

DB	OEBB	SBB	NR	EN 13803
1 : 600 Also applies to graduated transitions	New construction 1 : 800 Maintenance 1 : 600	2 ‰, i.e. 1 : 500	1 : 400	2.5 mm/m, i.e. 1 : 400

Fig. 3-22: Normal values for linear cant ramps [7] – [10] [19]

The length of the cant ramp can be easily calculated, taking into consideration the specified cant gradient.

Linear cant gradient	Schramm cant gradient	Bloss cant gradient
$l_R = \dfrac{d}{N}$	$l_R = \dfrac{2 \cdot d}{N}$	$l_R = \dfrac{3 \cdot d}{2 \cdot N}$

Fig. 3-23: Calculation of length of standard cant ramp based on cant gradient

l_R: length of cant ramp

N: cant gradient

Ideally, the length of the cant ramp should coincide with the length of the transition curve. Therefore, the lengths of both of these elements of track alignment are frequently matched to each other. Very often, the length of track that is canted dominates the length of both alignment elements. However, it can occasionally happen that the lengths do not match due to local circumstances. In Switzerland, for radii < 350 m, care has to be taken that the cant $ü_{red}$ at the end of the transition curve meets at least the following condition: [10]

$$d_{red} = \frac{11.8 \cdot v^2}{R} - 130 \qquad (3\text{-}25)$$

d_{red} in [mm]:   cant at the end of the transition curve

v in [km/h]:   speed

R in [m]:   radius

The length of track that is canted then has to be continued into the curve.

In contrast to this, in Austria, the end points of the transition curve and cant ramp must coincide. [20] Only the start may differ whereby the cant ramp must never be longer than the transition curve. If parts of the transition curve are designed without cant, the jolt or acceleration arising here must be compared with the limits.

In Germany and the UK, the transition curve and cant ramp should coincide. [8] [21] In general, the length required for the cant ramp is thus set, and the transition curve is adjusted accordingly.

Different lengths place special requirements on maintenance and may occur on existing lines. Their geometric design has to be implemented carefully, and the standard values specified must be precisely adhered to.

In addition to the transition curves mentioned above, standard EN 13803 also lists cosine and sine transitions. [7] Graduated transition curves are used cost-effectively in Europe when existing lines are to be prepared for higher speeds and no additional space is available for a new alignment. Existing space can frequently be better utilised with graduated transition curves.

In the UK, Network Rail permits the use of graduated transition curves, only when installation and maintenance equipment is capable of installing and maintaining the alignment. [22]

3.5.3   Cant deficiency and cant excess

The physical principles (see section 3.2) illustrate the effect of cant on the acting lateral acceleration which can be compensated completely by the calculated cant (equilibrium cant). Thus, theoretically, the passenger will not feel any lateral forces or acceleration in the curve. However, this applies only to the curve and not to the transition. Furthermore, a curve cannot be canted without the need to impose limits (Fig. 3-24).

Large cant values affect the standard clearance gauge due to the resulting inclination of the vehicle. This must be taken into account, especially at platforms. The cant could make disembarking more difficult, or the platform edge could present a fixed obstacle. Therefore, cant is frequently regulated in particular for these areas. Installing the equilibrium cant (equation (3-17)) compensates the lateral acceleration only for a specific radius and for a specific speed. However, on lines with mixed traffic trains travel at different speeds. If the cant is designed for fast passenger trains, the values will be particularly high. Slow trains will slide towards the inner rail due to the large lateral inclination, which increases the maintenance requirement for this rail. It is also to be expected that even fast trains must stop at signals or travel at reduced speeds.

In addition to a maximum cant, a minimum cant is also defined in the DACH countries. These values are 20 mm in Germany [8] and Austria [16]. In Switzerland, they have been set at 10 mm [10]. There is no specified minimum in the UK.

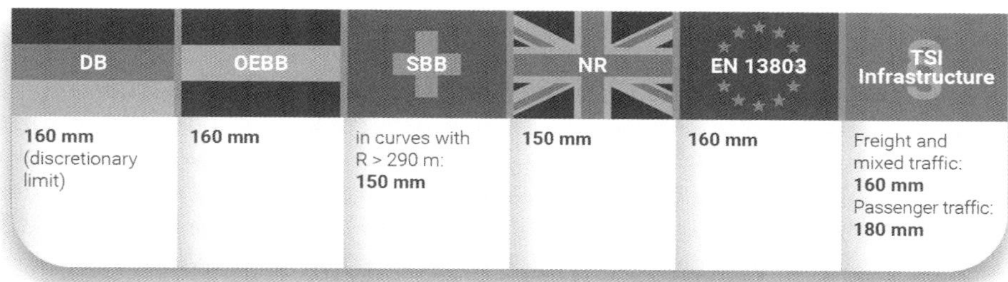

Fig. 3-24: Maximum cant to be applied [7] – [10] [23]

Generally, for all railways, the cant applied to curves is a reduction of the equilibrium cant and is, in fact, a compromise. The cant selected, with engineering judgement, will be too large for slow trains and too small for fast ones.

$$\Delta d = d_{EQ} - d = 11.8 \cdot \frac{v^2}{R} - d \qquad (3\text{-}26)$$

v in [km/h]:   train speed

R in [m]:   radius of curve

d in [mm]:   applied cant

d_{EQ} in [mm]:   equilibrium cant

The value for Δd is positive or negative depending on the inputs. Accordingly, this is called either a cant deficiency (positive) or a cant excess (negative). The values are limited by the speed, after taking into account the maximum lateral acceleration (ride comfort and safety of goods). The country-specific guidelines implement this condition differently. In Germany, the standard cant as per RIL 800.0110 [8] is calculated as follows:

$$d = \frac{6.5 \cdot v^2}{R} \qquad (3\text{-}27)$$

d in [mm]:   standard cant

R in [m]:   radius

v in [km/h]:   speed

In contrast to the actual equation of equilibrium cant (3-17), the value of 11.8 is reduced by 45 % to 6.5. This results in a deficiency which has to be checked against the applicable discretionary limit. [24, p. 37] The Swiss guideline deals with this topic in a similar manner, although R I-22046 also distinguishes between different speed ranges and train configurations. [10] Thus, the equation (3-27) changes at speeds from 160 km/h up to and including 200 km/h as follows:

$$d = \frac{6.2 \cdot v^2}{R} \tag{3-28}$$

d in [mm]:   standard cant

R in [m]:   radius

v in [km/h]: speed

In Austria, the cant deficiency/excess is integrated into the calculation to a much greater extent and thus results in different values for the permissible cant, depending on radius and speed. The cant deficiency in track is limited to 100 mm for new construction and maintenance of TEN lines. The recommended limit for cant excess is specified in the rules and standards as 80 mm for new construction and 110 mm for maintenance of track for TEN lines.

In the UK Network Rail sets a cant deficiency limit of 110 mm for new plain line track. [25]

3.6   Minimum lengths

In addition to the absolute values for lateral acceleration, its change over distance is of major importance for ride quality. This is a fact which already plays a major role in the smoothing of transition areas between elements of the alignment geometry. A rapid change in acceleration would impair the ride quality noticeably so that a minimum length is specified for individual elements of the alignment in the DACH countries. The values shown in Fig. 3-25 are guide values for curves and straights. In the UK, the minimum length of each geometrical element shall be appropriate to the length and characteristics of vehicles likely to use the track. [26] Network Rail requires the number of individual elements (straights or curves) to be kept to a minimum and that each element should be as long as possible, and not normally be of a length equal to less than 2 seconds at a maximum line speed. It does, however, make specific requirements for the minimum length of track geometry elements in transition curves. (see Fig. 3-25) [27]

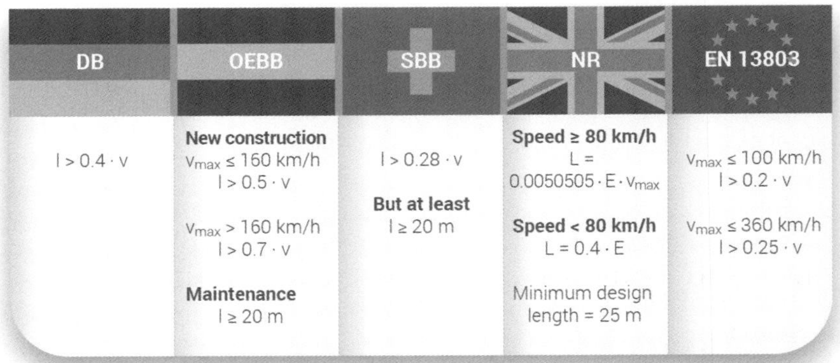

DB	OEBB	SBB	NR	EN 13803
$l > 0.4 \cdot v$	**New construction** $v_{max} \le 160$ km/h $l > 0.5 \cdot v$ $v_{max} > 160$ km/h $l > 0.7 \cdot v$ **Maintenance** $l \ge 20$ m	$l > 0.28 \cdot v$ **But at least** $l \ge 20$ m	**Speed $\ge$ 80 km/h** $L =$ $0.0050505 \cdot E \cdot v_{max}$ **Speed < 80 km/h** $L = 0.4 \cdot E$ Minimum design length = 25 m	$v_{max} \le 100$ km/h $l > 0.2 \cdot v$ $v_{max} \le 360$ km/h $l > 0.25 \cdot v$

Fig. 3-25: Examples for the calculation of the minimum lengths of track geometry elements for curves and straights in the DACH countries [7] – [10] [27]

3.7 Reverse curves with an abrupt change in curvature ("Track Scissors")

In practice, it may not be possible to arrange the required intermediate elements or implement the required minimum lengths. In restricted conditions, for example, narrow valleys, along rivers or on mountain routes, the options are highly limited (Fig. 3-26). It may, therefore, be unavoidable to have two reverse curves with transition curves passing directly into each other.

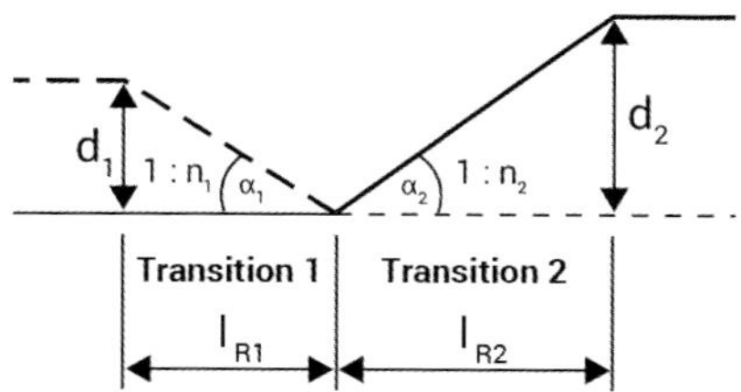

Fig. 3-26: Direct sequence of two canted ramps in a reverse curve

In principle, elements of the alignment with a reverse curve (curve or transition curve) should not adjoin each other. However, if this is necessary, "track scissors" can help. "Track scissors" can be arranged where the canted transition of one reverse curve abuts the canted transition of the preceding curve or where the minimum straight cannot be placed between them.

"Track scissors" have the objective of rotating or transitioning the vehicle continuously from one tilted position into the opposite tilted position in the reverse curve. Schramm [6, p. 18] indicates two options for this, in addition to the required intermediate straight.

– creation of a new canted ramp by lowering the originally canted rail (variant A);
– lifting and lowering of the opposite outer rails of both curves (variant B) with the same gradient.

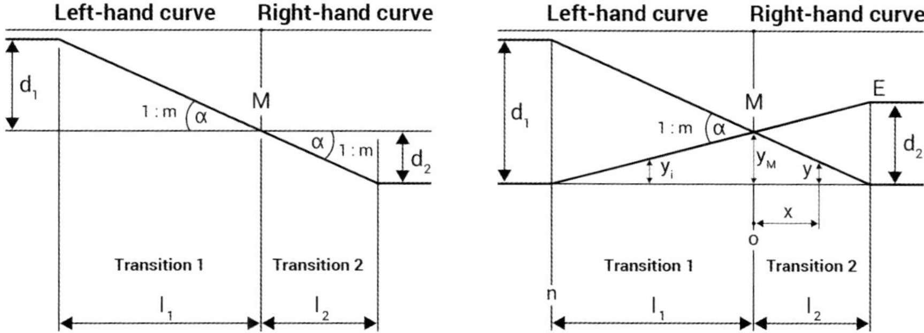

Fig. 3-27: Different options of "track scissors", variant A (left) and variant B (right) [6, p. 18]

In the DACH countries, only variant B is approved. Its dimensioning is mainly a result of the cant values of both curves. Each of the transition curves ends at point M, while the ramp continues linearly into the curve.

$$y = \frac{y_m}{l_n} \cdot (l_n - x) \tag{3-29}$$

$$\text{with: } y_M = d_1 \cdot \frac{l_1}{l_1 + l_2} = d_2 \cdot \frac{l_2}{l_1 + l_2} \tag{3-30}$$

"Track scissors" are only to be used with linear cant ramps. In Germany, it is permissible to arrange graduated transitions one after another. The use of "track scissors" is necessary in Austria, despite the smoothing of the inflexion points.

The use of short curves increases the flexibility in finding a trace. However, if the length of the curve is less than the threshold value, it is not permissible to use this as the ride quality is reduced. The use of graduated transitions makes an alternative design possible in this case: the vertex curve. For this, two transition curves will have the same length and their cant ramps follow each other. This is possible only with graduated transition curves and is, therefore, used mainly in Germany.

3.8 Buffer locking

Alignment design without transition curves should be avoided as a matter of principle. If nonetheless two curves of opposite orientation meet, safety-relevant criteria, as well as aspects of ride quality are of importance.

TSI Infrastructure pays particular attention to reverse curves with a radius between 150 and 300 m [13, section 4.2.3.4]. An adequate straight section is required between these to avoid possible buffer locking (Fig. 3-28). This means the possible jamming of the buffers of two consecutive vehicles. If such locking is not released, the forces between the vehicles will result in a derailment. To prevent buffer locking, a straight section, which can be determined as a function of the radii of the two curves, has to be placed between the reverse curves.

In the UK, Network Rail identifies a particular track design case where, with a standard spacing between tracks of 1970 mm, there is a risk of buffer locking or extreme lateral movement of corridor connections especially with crossovers using turnouts with smaller radius switches. [28]

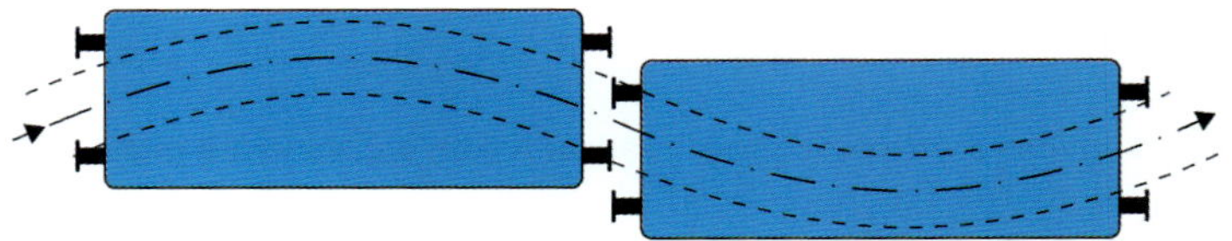

Fig. 3-28: Schematic diagram of buffer locking in a reverse curve

3.9 Track re-alignment – variation of the distance between the centres of two tracks

The distance between track centres ensures the space required between two tracks so that both can be used for traffic at the same time. However, it may be necessary to adjust this distance before stations or groups of turnouts. In this case, it will be increased at these points and afterwards reduced back to the standard distance. Such changes of the distance between tracks are achieved by design changes to the basic track layout. The distance between two parallel tracks can easily be changed, e.g. by a different choice of radii or transition curves of different lengths (change in circular curves). However, if these changes are not possible through the design of the basic layout, the principle of parallel track re-alignment is used in the DACH countries.

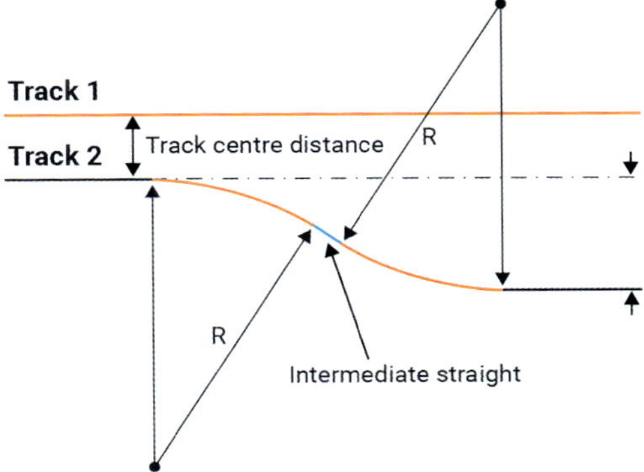

Fig. 3-29: Track re-alignment with two circular curves with intermediate straight without transition curve or cant ramp

In general, it is possible to distinguish between three different types of parallel track re-alignment: [6, pp. 61ff]

– track re-alignment with two circular curves without cant, without transition curve and intermediate straight (rare)
– track re-alignment with two circular curves without cant, without transition curve and with intermediate straight
– track re-alignment with two circular curves with cant, with transition curve and without intermediate straight.

Track re-alignment with cant and transition curves are classic "track scissors" and are laid out by the principles described. If no transition elements are used, the basic alignment limits must be taken into account. RIL 800.0110 requires the use of curves without transition curves and cant ramps for track re-alignment up to 1.5 m. [8] These curves must have a radius of at least:

$$R \geq \frac{v^2}{2} \qquad (3\text{-}31)$$

The minimum lengths of straights and curves must be taken into account. If no intermediate straight is used, the possibility of buffer locking due to the reverse curve must be investigated separately.

In the UK the most common example of reverse curves with a minimum length of straight track between the curves are turnouts with track reversing to a parallel main line. In order that the design can both minimise the lateral acceleration and prevent buffer locking, Network Rail specifies a maximum speed for each geometric design of turnout. [29]

3.10   Gradient change

Railway lines inevitably have to traverse differences in height across a terrain. However, due to the specific characteristics of the wheel/rail contact, there are constraints on possible longitudinal gradients. The low friction between rail and wheel is advantageous for the transport of large masses but also has a few disadvantages. If the gradient is too large, long heavy goods trains cannot start without support as the wheel cannot transfer sufficient energy to the rail due to the low friction coefficient.

In stations or on track used for the formation or stabling of trains there are additional constraints on the gradient to prevent the wagons from rolling away. However, this cannot be excluded completely; for this reason, vehicles must always be secured against rolling away. [18, p. 609]

Longitudinal gradient:

Germany: s   Austria: Θ   Switzerland: I   UK: no specific symbol

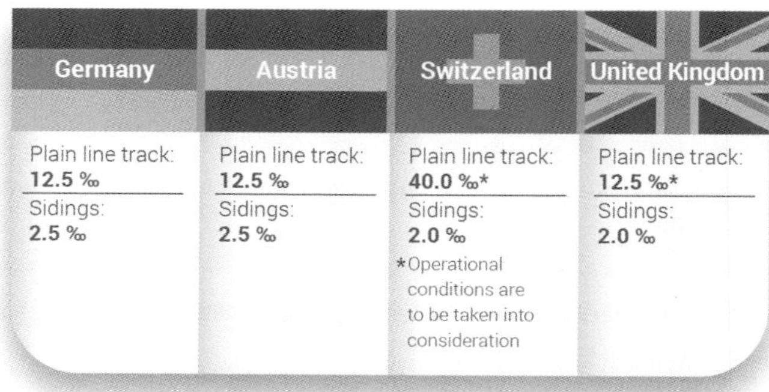

Fig. 3-30: Maximum gradient for new construction [1, Art. 7(1) and (2)], [2, AB 17(7.1.4 and 7.2)], [3, Art. 14(2) and (3)] [30]

If it is not possible to route the line through the terrain, taking into account these maximum gradients (Fig. 3-30), engineering structures, such as cuttings or tunnels, are required. As a new line is constructed, the soils arising from these types of earthworks can be used as "fill" to smooth differences in the height of the terrain (see chapter 2). However, there are always changes in a gradient over the length of a line. These occur where two line sections with different gradients meet. Although the focus so far has been on the design of horizontal alignment, similar principles can be applied to the vertical trace. Here, too, changes in the gradient of the line can be perceived as a vertical jolt. This is limited in the design of railway in the same way as the horizontal jolt.

These changes in gradient are smoothed (Fig. 3-31) in Germany [8] and Austria [16] if the difference between the gradients meeting (Δp) is greater than 1 ‰. In Switzerland, changes in the gradient of less than 2 ‰ are smoothed similarly with a kink. [10] A fictitious radius is chosen for the kink, which implements a vertical curve over a length l of 0.5 to 1 m.

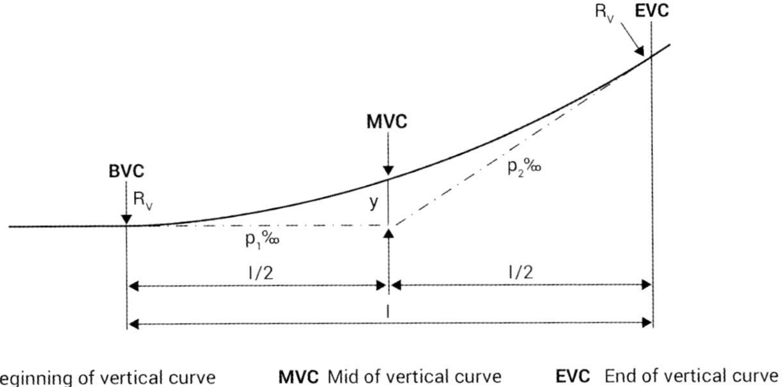

BVC Beginning of vertical curve   **MVC** Mid of vertical curve   **EVC** End of vertical curve

Fig. 3-31: Diagram of a change in vertical gradient and its main geometric points

EN 13803 provides for either a conventional circular curve or a 2nd degree parabola for this smoothing. [7, section 5.2] The standard is slightly more precise than the railway companies' rules and standards when stating those limits. It also distinguishes between speed ranges for the mandatory vertical curve:

$0 < v \leq 230$ km/h   1 ‰

$230 < v \leq 360$ km/h   0.5 ‰

In principle, gradient changes are to be avoided on bridges and in switches and crossings as construction and maintenance are particularly difficult here. Ideally, changes in gradient are provided for on the straight since complicated geometries can arise in a curve due to the combination with the cant.

Depending on the type of vertical curve, a distinction is made between valleys ($p_1 < p_2$) and humps ($p_1 > p_2$). The radii, as calculated in Fig. 3-32, are used for the vertical curve itself. The values for horizontal curves are, however, much greater in comparison. The reason for this is that vertical acceleration up to 0.31 m/s² is regarded as acceptable. This value is only one third of the permissible horizontal lateral acceleration of 0.98 m/s²; this is why the values for horizontal curves are much greater. In the UK, Network Rail's normal design value for vertical acceleration is 2.25 % g. [31]

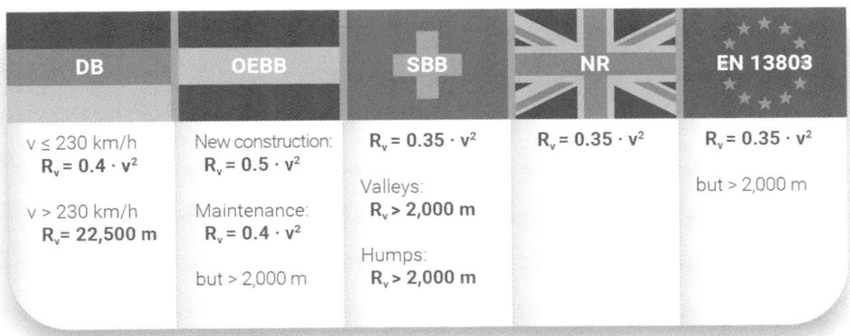

Fig. 3-32: Examples of the limits of vertical curve radius Rv [7] – [10] [32]

Based on the curve radius determined, the following equations result for the length l or $\dfrac{l}{2}$

$$\frac{l}{2} = \frac{|p_1| - |p_2|}{2000} \cdot R_v \qquad (3\text{-}32)$$

and the distance between peaks

$$y = \frac{1000 \cdot l^2}{8 \cdot R_v} \qquad (3\text{-}33)$$

p_1 and p_2 in [‰]: gradients

R_v in [m]: vertical curve radius

l in [m]: horizontal length of curve

y in [m]: distance between peaks

Referring to Fig. 3-31 above and using these values, the vertical curves can be uniquely described, and the main points start of vertical curve (BVC), centre of vertical curve (MVC) and end of vertical curve (EVC) can be determined. Unlike for horizontal curves, transition curves are not necessary in vertical curves.

3.11 Summary

In this chapter we have looked at the basics of plain line track design. Today, the majority of track design is undertaken using computers and specially designed software. Even so, as this method of track design does not rely on the user using empirical formulae, the resulting trace has to demonstrate compliance with national and railway standards for passenger comfort, the transport of goods and above all safety of the railway. It is also important that the available means to install and maintain the permanent way has the ability to maintain the geometric calculations used in the design. We shall start to look at this in the next chapter.

4 Track geometry defects and the determination of correction values

Wolfgang Nemetz and Richard Spoors

4.1 Key issues

This chapter deals with the appearance and impact of track geometry defects and the determination of their correction values. These form the basis for the tamping process, which will be dealt with in chapter 5. We will explain the essential technologies and provide answers to the following issues:

- What are track geometry defects, how do they occur and what are their effects?
- What maintenance processes can be used to rectify track geometry defects and how do they differ?
- What is the difference between absolute measurement and relative measurement? Which method can be used for what?
- What is the precision method and in what way does it differ from the compensation method?
- Which measurement methods are suitable for surveying, pre-measuring and post-measuring?

4.2 Track geometry defects

4.2.1 Definition and their formation

Track geometry defects are deviations of the position or level of a track from its planned longitudinal level or alignment. The latter is a result of the design of the track and is defined as the target track geometry.

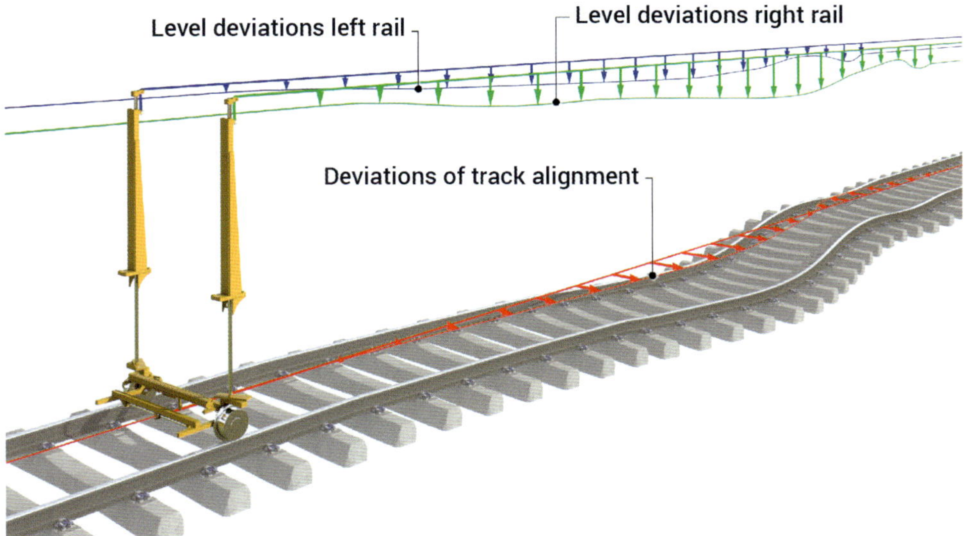

Fig. 4-1: Track geometry defect diagram

Track geometry defects occur from the use of the track under operational influences as a result of deformations of the structural system consisting of the track, substructure and subsoil as well as a result of an ineffective track drainage system and climatic influences. Their formation is encouraged by inhomogeneous track progressions. These can include different elasticities in the track and substructure in transitions, such as a change of sleeper type, bridges without a continuous ballast bed, slab track, level crossings, different ballast bed thicknesses, etc. The track geometry defect will intensify further as a result of the impact of dynamic forces when travelling over the defective spot.

Fig. 4-2: Track geometry defect [1]

4.2.2   Distinction of track geometry defects

EN 13848 [2], the standard for track geometry quality, stipulates that track geometry defects have to be corrected once specific values have been reached. It defines track geometry defects for the following parameters:

Longitudinal Level, Alignment, Twist and Track Gauge
Longitudinal level and alignment are typically measured with inertia measurement systems and the raw data is transformed and bandpass filtered into wavelengths of interest. EN13848 stipulates them as:
- D1 (short waves), wavelength $3 \leq 25$ m
- D2 (medium waves), wavelength $> 25 \leq 70$ m
- D3 (long waves), wavelength $> 70 \leq 150$ m for longitudinal level and $> 70 \leq 200$ m for alignment

The wavelengths of interest (specifically their limits) are aligned to the line speed. For the slower speeds used predominantly by commuter trains (v< 160 kph) the wavelength is in the range D1, for speeds between 160 kph and 230 kph (Inter City trains) the wavelength range is extended by the inclusion of D2. For the highest speeds above 230 kph the wavelength range is further extended by the inclusion of D3.

The UK traditional practice is to utilise high pass filters set at wavelengths of interest as follows:
- 35 m (short waves), wavelength $0.5 \leq 35$ m
- 70 m (medium waves), wavelength $0.5 \leq 70$ m

These UK wavelengths for longitudinal level (top) and alignment output profiles are specified in standard RT/CE/S/042 – Track Geometry Recording and the use of a 35 metre wavelength with a measuring capability of 40 mm is different to the DACH countries and the European standard that use a shorter 25 metre wavelength. Therefore, the resulting outputs as standard deviations may not be directly comparable.

EN 13848 makes provision for either an assymetric (preferable) versine measurement system or an inertia measurement system. For the versine measuring system the output is transformed into the equivalent D1, D2, D3 by the decolouring process so that the measurements can be directly compared with D1, D2 and D3 limit values.

- **Longitudinal level**:
  Here, the progression of the track centre line is shown in an elevation view.
  Short and medium waves of the longitudinal level mostly arise from the interaction between the track and the vehicle (depending on the speed and the condition of the vehicle). The causes here are primarily found in the ballast bed and the substructure, for example, wet spots, deteriorated ballast, settlements in the substructure which may be caused by drainage damage, etc.
  On the other hand, the trigger of long waves can be found in the substructure or subsoil. [3, p. 1] While long waves generally have no impact on the immediate operational safety, short and medium waves may have unwanted consequences, leading to derailments in the worst case. This is due to the increase in wheel acceleration and with it the vehicle body acceleration. Waves that are shorter than 3 m are generally caused by rail defects (e.g. rail joints, insulated joints, etc.) and cannot usually be repaired by tamping.
- **Alignment**:
  The progression of the track centreline is shown in plan view.
- **Twist**:
  This is defined as the change of the cross level in relation to the distance apart of the measurement points. The standard defines twist as follows:
  "The algebraic difference between two cross levels taken at a defined distance apart, usually expressed as a gradient between the two points of measurement." [2, section 4.6.1] Track twist defects will generally arise from one rail sinking on a short section but can also arise from the effects of frost (frost lifts) or heat (vertical track deformation), i.e. from the track "edging up" on one side.
- **Cant (cross level)**:
  A cant is the difference of level between the left and right rail. It is defined in the standard as follows:
  "The cant is determined by measuring the difference in the height of the adjacent rail running surfaces." [2, section 4.5.2]
- **Track gauge**:
  The smallest distance between the rails is the track gauge. DB and ÖBB measure between 0 and 14 mm below the top of the rail, SBB precisely at 14 mm below the top of the rail. The standard defines gauge as follows:
  "Track gauge, G, is the smallest distance between lines perpendicular to the running surface intersecting each rail head profile at point P in a range from 0 to Zp below the running surface. Zp is always 14 mm." [2, section 4.2.1]
  Network Rail defines track gauge as the distance between the running edges of the rails in a track, measured without load, at right angles to the running edges of the rails in a plane 14 mm below their top surface. [4]
  It should be noted that any defects in track gauge cannot be corrected by the tamping process. This should be especially considered when track quality indices that combine different measurement signals are used to plan maintenance work.

4.2.3   The Principal Measurement of Track geometry faults in the UK

Fig. 4-3: Network Rail's New Measurement Train is able to record track geometry at speeds up to 200 kph. [11]

Network Rail operates a number of track condition monitoring trains, including the New Measurement Train (see Fig. 4-3) which between them record and report on the 20,000 track miles of the UK network. There are five main types of track geometry fault that Network Rail's measurement trains help to prevent in the management of derailment risk. These are twist, gauge, cyclic top (see below) and horizontal and vertical alignment (longitudinal level). Network Rail's standard "Track geometry – Inspections and minimum actions" defines thresholds for the first two of these parameters that include blocking the line, adopting the limits defined in the Railway Group Standard – "Track System Requirements". For severe cyclic top there are threshold limits for the imposition of speed restrictions. All five parameters have threshold limits for AI, IL and IAL with prescribed timescales for fault rectification.

Many of the measurement principles used today have their foundation in development work done by British Rail Research in the 1960s. The introduction of the Matisa track recording trolley in 1957 was followed a few years later by the mechanical recordings being converted with an early on-board computer into numerical weighted outputs for twist, top and line in six bands of exceedance per quarter mile. This was translated into punched tapes that ran on mainframes from which print-outs were produced as an aid to Track Engineers planning their maintenance work. [5]

Later in the 1960s, British Rail's Research organisation at Derby fitted an early inertial measuring system to a diesel multiple unit called "Laboratory 5" and started to measure track geometry with electronic sensors. This led a few years later to the fitting of similar contactless equipment to a Mark IIF passenger vehicle that entered service in 1975 with an operating speed of 110 mph (175 kph) called the High Speed Recording Coach (HSTRC). In parallel with the development of hardware was the analysis of data to produce information. The new inertial measuring system was based on the loaded profile of the wheel at the point of contact and optical sensors enabled measurements to be made in the horizontal plane. Electrical high-frequency-pass filters are used to remove the long wavelength and design information, effectively creating a reference line based on a moving data average. BR decided upon two longitudinal wavelength cut-offs for data analysis; 35 metres for suburban and cross country routes supplemented by 70 metres for high speed lines. These values continue in use today.

Today, track quality indices are presented as Standard Deviations (SDs) measured over one eighth of a mile (roughly 200 m). Standard deviation is the measure of the amount of variation of a random process from an assumed mean, therefore, as track vertical and horizontal profile data have sufficient similarity to random processes, a measure of their magnitude can reliably be made in this way. The standard deviation may be thought of as a measure of the energy imparted to a vehicle suspension due to the roughness of the track. A value of 0 corresponds to perfect track and a value of 9.9 mm to extremely poor quality. Therefore, a scale of intermediate values can be produced to specify the quality of all classes of line. Each SD gives a measure of the extent to which the track deviates from the ideal horizontal or vertical geometry for the particular wavelength to which it relates over the longitudinal imperial distance of 1/8th mile. [6] Railway Group Standard "Track System Requirements" defines limiting SD values for top and line by speed band as either very poor or maximum above which action is required to restore track geometry or impose a speed restriction. (See Chapter 2 Fig. 2-14). Network Rail incorporates these definitions in a Track Geometry Quality Band table in its standard "Track geometry – Inspections and minimum actions", where the band descriptions are Good, Satisfactory, Poor, Very Poor and Maximum. [7] Any values meeting or exceeding the maximum band are known as "Super Red". Although the SD value is a measure of geometry quality rather than a defect value, it is a valuable tool in the management of track at ground level and also from a strategic perspective, where Network Rail provides two high level measures of track quality as performance indicators. Here, over a large geographic area, the number of eighths of a mile of track in the good and satisfactory bands as a percentage of the total is shown as Good Track Geometry (GTG). This may be compared with a similar calculation for all the track in the very poor band (Poor Track Geometry). At a strategic level over time these measures can be useful in showing trends in the management of the track asset.

4.2.4   The Phenomenon of Cyclic Top

Cyclic top is a particular track geometry defect that is difficult to recognise and has caused a number of freight train derailments in the UK over the last 35 years. Cyclic top causes resonance in a train's suspension system. If the resonance reaches severe levels, the train can derail by the flange wheel lifting vertically above the rail head. Train suspension types that are susceptible to this are in freight vehicles as they have no designed dampening of the suspension response, unlike passenger vehicles. Cyclic top occurs when a dip in a rail causes the suspension of a vehicle passing over it to bounce. The track at the end of the bounce then receives an impact loading that can create a second rail dip and, as that second dip deteriorates with successive impacts, trains will then bounce a second time, creating a third dip. After time a sequence of dips can be created over a particular wavelength which will, depending on the suspension characteristics of the vehicles using the line, cause each successive bounce to increase to the point where the vehicle will derail. Typical wavelengths experienced in the UK are 4.5 m, 6 m, 9 m, 13 m and 18 m.

The problem occurs in both jointed track and track with continuously welded rail (CWR). Experience has shown that a number of track features may lead to cyclic top defects. These include dipped joints in jointed track, dipped welds in CWR, weak formation and poor geometry. Routes with regular flows of particular freight vehicles with two axles (but not exclusively) are particularly at risk.

Research into derailments where cyclic top was found to be the cause has resulted in the design of computer software filters that have been installed on Network Rail's fleet of Track Recording Vehicles (TRV) (see below). Cyclic top is not visible on the longitudinal level graph for either D1 or UK Top 35 and can only be detected by computation. It can be detected in several ways and the

UK method is to apply a band-pass filter to the longitudinal level signal around each wavelength of interest for both rails. An algorithm is then set to count the number of and the severity of the peaks above a threshold to simulate the resonant energy build up in the suspension. Limit values are set for increasing severities of this simulated resonant energy. Derailment risk can increase dramatically when lower severity cyclic top is combined with twist or alignment isolated defects. The UK track recording vehicles are programmed to look for and report such combinations.

In the UK, when cyclic top has been discovered by the track recording vehicle, a report is produced showing the length of track over which cyclic top faults at the alert level and intervention level have been found. These are defined in standard NR/L2/TRK/001 MOD11 at Appendix A. Where peaks meet the immediate action level (or critical limit value) the mitigation of the risk to derailment is to impose a speed restriction of 30 mph (50 kph) to all freight vehicles. This prevents resonance occurring at various critical speeds until the actual dips can be effectively repaired.

The TRV cyclic top report shows the GPS position, wavelength, number of cycles of the cyclic top fault and the category of severity. Track Maintenance Engineers can then use this to take local action and instigate prevention measures to manage the safety of their track. A surveying trolley has also been developed in the UK specifically to measure cyclic top (see Fig. 4-49) and this can be used on site in conjunction with the TRV report to determine more accurately geometry faults and the necessary intervention. [8]

4.2.5   Intervention limits

In the European Union, railway infrastructure managers must observe the applicable EU standards and thus the TSI standards (Technical Specifications for Interoperability). These ensure a minimum standard of the technical equipment. These regulations also stipulate that a maintenance plan has to be drawn up by the infrastructure managers. This maintenance plan must regulate the required inspections and their frequency, the staff qualifications as well as the measurement methods. In addition, the infrastructure manager has to define immediate action limits for safety-relevant parameters and subsequent actions if these limits are exceeded. [9, p. 41].

As the standard requires measurements on the loaded track, inspection measurements are usually carried out with the use of track recording vehicles. Track recording vehicles are self-propelled or hauled vehicles with permanently installed recording equipment and systems for measuring, evaluation and recording of track geometry parameters under load that provide measurement results in line with the requirements of EN 13848-1 [2].

EN 13848-5 defines three primary intervention and alert limits (see also Fig. 4-4): [10]

– Immediate action limit (IAL): If a certain value is exceeded, this requires immediate actions that reduce the risk of derailment to an acceptable level. This can be done by either closing the track, reducing the speed or correcting the track geometry.
– Intervention limit (IL): If a certain value is exceeded, corrective maintenance activities are required so that the immediate action limit cannot be reached before the next inspection.
– Alert limit (AL): Exceeding a certain value requires an analysis of the track geometry condition. This condition must be taken into consideration in the scheduled maintenance activities.

The United Kingdom has left the European Union, however, the standards framework and relevant requirements will continue as they are whether derived from the EU or UK. [12] Network Rail specifies their primary intervention limits for track in their Track Standard NR/L2/TRK/001/MOD 11 "Track geometry – Inspections and minimum actions".

	EN 13803 Part 5 [5]	DB	ÖBB	SBB	Network Rail
Alert limit	AL	SR_A	AS	-	AL
Intervention limit	IL	SR_{100}	ES	ES	IL
Immediate action limit	IAL	SR_{Lim}	SES	SES	IAL

Fig. 4-4: Designation of the intervention limits in the DACH countries and the UK [13]

AL	Alert limit
IL	Intervention limit
IAL	Immediate action limit
SR_A	Fault response A: Plan repair activity
SR_{100}	Fault response 100: Repair by next inspection
SR_{Lim}	Fault response Lim: Adverse effect to be expected
AS	Aufmerksamkeitsschwelle (alert limit)
ES	Eingriffsschwelle (intervention limit)
SES	Soforteingriffsschwelle (immediate action limit)

Fig. 4-5: Explanation of intervention limits

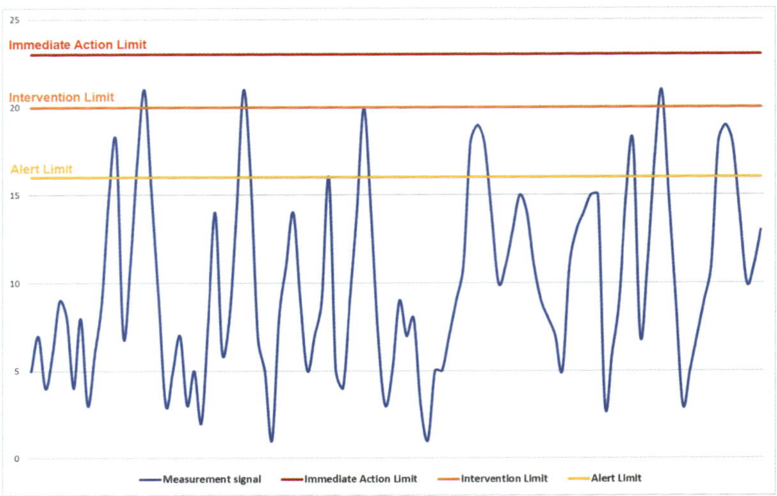

Fig. 4-6: Schematic diagram of intervention limits

In this fictitious example, for the purpose of visualising the intervention and alert limits, the location was entered in the x-axis and the measurement values in the y-axis. The alert limit was exceeded in five instances, the intervention limit in three instances. Consequently, the infrastructure manager would have to take appropriate action in line with their relevant regulations.

The railway administrations have different systems for monitoring the observance of the limit values. Depending on their design, they can record and evaluate the results of measuring runs, document limit exceedances and, based on the history of the track geometry parameters, may even be able to forecast the future development of the track geometry. Based on this, the most appropriate action from an economical and technical viewpoint can be taken.

4.2.6 Maintenance methods for correcting track geometry defects

Track geometry defects can be corrected by the following processes:
- pre-planned maintenance tamping
  - DB: mechanised maintenance (Durcharbeitung DUA)
  - ÖBB: predictive maintenance tamping
  - SBB: maintenance tamping
  - NR: maintenance tamping or stoneblowing
- isolated defect correction

In the isolated defect correction (spot tamping), unexpected track geometry defects of a short length (approx. 3 to 20 m) are corrected. This is essentially considered a temporary measure. In most cases, isolated defects have to be corrected at short notice and, if no machine is available, the intervention is undertaken manually as a short term solution to ensure safety of the line or a speed restriction may have to be imposed.

In contrast, the mechanised maintenance (or preventative maintenance) is considered to be planned, sustainable maintenance. This mechanised maintenance is subject to medium-term planning with lead times of between 0.5 and 2 years (depending on the railway administration).

4.3   Basic principles of measuring track geometry

The quality of the track geometry that can be achieved with a tamping machine depends on the quality of the correction values available to the machine. These values can be determined by two different strategic approaches:

- the precision method
- the compensation method

A defined track geometry can be achieved with the precision method, whereas the compensation method can only achieve a proportional reduction of the track geometry defects. This is because to accurately determine the correction values, the difference between the target geometry (the through alignment design) and the actual track geometry has to be measured. This requires that the target track geometry be known.

The measuring process is generally divided into three steps:

1. Survey: This is usually a geodetic survey for calculating a route layout. On existing tracks, the geometry of the track in the field as well as constraints and fixed points (marking points) are also recorded.
2. Pre-measuring: Here, the geometry of the existing track is measured in order to determine the difference between the target and actual track geometry and obtain the lifting and slueing values for the subsequent correction of the track geometry.
3. Post-measuring: After completion of the geometry correction, post-measuring will be carried out to document the correct execution of the work. Furthermore, it will also serve as the basis for reopening the track for operation at the design line speed.

The railways distinguish between absolute and relative measurements which also differ with regard to their reproducibility.

Absolute measurement is also known as coordinative or geodetic measurement. Here, the position of the rail in the field is recorded and mapped in classification systems with associated coordinate systems.

Modern geodesy is based on the premise that the earth is displayed as an ellipsoid or geoid. The geoid consists of irregular surfaces of equal gravitational potential that cuts the perpendiculars of gravity at a right angle throughout its surfaces. The positional reference surfaces represent ellipsoids that can be described mathematically. This results in different reference or classification systems.

Classification systems (so-called reference systems) are physically defined identification systems. They enable the spatial allocation of information to each other and are mostly defined specifically to countries. The reference systems are anchored by reference points (fixed points).

Following harmonisation and the use of new technologies, such as satellite-based surveying, new global reference systems (e. g. WGS 84 or ETRS89) were introduced.

Coordinate systems show the mathematical mapping rules for describing the position of points in space. Within a reference system, different coordinate systems can be converted as needed.

The coordinative measuring/route layout generates absolute coordinates relating to a given reference system of both the actual data and the target data from the route layout. This process is divided into the following steps:

- geodetic survey
- route layout
- drawing up plans (position, pegging, level plans)
- data archiving
- pegging out (= transfer of a drawing into the field)

The geodetic survey or pegging out is carried out with instruments such as total station, theodolite, tachymeter, levelling device, GPS receiver, etc.

Figure 4-7: Total station instrument

A track centreline that has been defined using coordinates can be reproduced at any time, even if marking or fixed points, such as mast bolts, are lost or changed.

Relative measurement records only the track geometry. It is normally determined by adding up deviations. A reference to its position in the field is generally not established. When determining the track geometry on so-called fixed points which are mapped in coordinates and which also serve as the end points of the measuring chord, a relative measurement becomes an approximate absolute measurement (= marked tracks).

Relative measurements are procedures based on versines. These are simple, yet accurate measurement procedures which (as a complementary system combined with geodetically recorded fixed points) can be used for the implementation of track and turnout construction work (maintenance, modifications, renewals) at ÖBB and SBB.

As of 01.01.2020, DB as a rule only permits geodetic surveys/pre-measurements in its own DB_REF classification system (see Fig. 4-8).

Fig. 4-8 shows the measurement methods that are to be used in line with the rules and standards of the respective DACH countries.

	DB	ÖBB	SBB	NR
Measuring/ Survey	Mandatory from 01.01.2020: geodetic measurement in DB_REF fixed point network. Track marking points and track checkpoints are to be integrated in DB_REF network.	Geodetic measurement, fixed points (e.g. masts) are to be integrated. Relative measurement can be used in exceptions.	Geodetic measurement (Toporail), fixed points (e.g. masts) are to be integrated.	New works, renewals and ATG require Geodetic surveys. Relative measurement for all other maintenance tamping.
Pre-measuring	Mandatory from 01.01.2020: geodetic pre-measuring, track marking points and track checkpoints are change points. Until then (and in exceptions): long chord measurements according to route layout.	Long chord measurements relating to fixed points (masts).	Pre-measuring not required if using PALAS. During work the tamping machine will use geodetic pegging as reference. If PALAS cannot be used: long chord measurements.	Pre-measuring by tamping machine, except for ATG.
Quality control Post-measuring internal track geometry	Measurement record as per EN 13231-1 and EN 13848-3.	Measurement record as per EN 13231-1 and EN 13848-3. For new/renewed track additional post-measuring with long chord measurement with integration of fixed points.	Measurement record as per EN 13231-1 and EN 13848-3.	Geodetic survey, NMT (NR New Measurement Train) or for maintenance, measurement run by tamping machine.

Fig. 4-8: Measurement methods permitted by the railway administrations of the DACH countries [14], [15] and the UK [16]

GOLDSCHMIDT
Smart Rail Solutions

When **measuring tracks and turnouts,** there are two options for reconstructing or re-establishing an originally calculated target geometry:

- the geodetic pegging out and
- marking the target geometry onto marking points.

When setting out the target geometry, it is projected to the designated marking or fixed points (with the use of lateral distance and offset measurement). These are put on catenary supports (if available). Special marking point supports have to be used on non-electrified lines, e.g. rail stayers, bridge railings, etc.

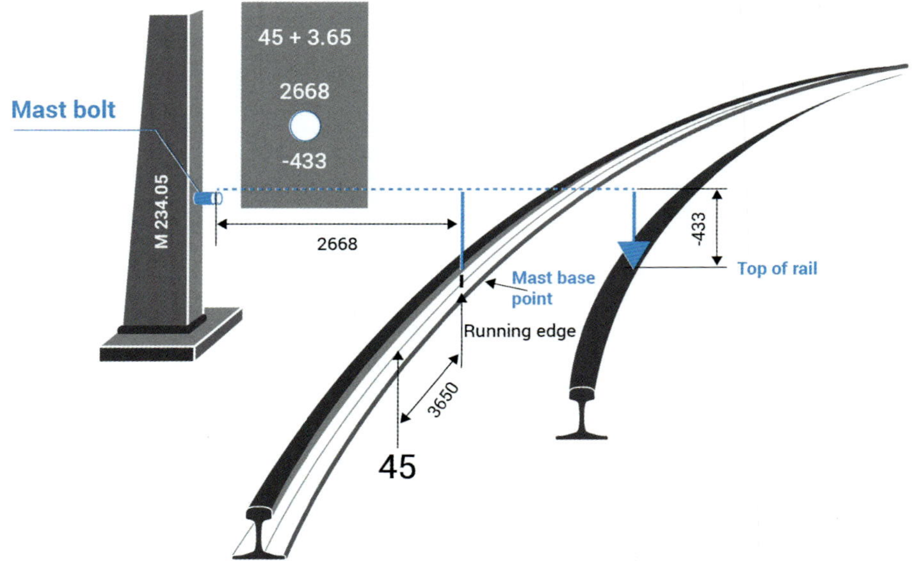

Fig. 4-9: Marking diagram (example ÖBB)

4.4 Precision method

With the precision method, the track is measured in line with geodetic principles with reference to the fixed points, and a calculation is made of the correction values from the difference to the target (design) geometry.

In order to position the track as required using the control systems of the tamping machines, data on the track geometry in plan view (alignment) and elevation view (level) as well as on the cant are required in addition to the lifting and slueing values.

In the UK, Network Rail has been using this method on the fast lines of the West Coast Main Line between London and Glasgow since 2004 to support the safety case for running of Pendolino and Voyager tilting trains. Precision method is also used at specific sections of other routes that have been designated Managed Track Position. This new concept was introduced in 2020 with a new Standard (NR/L3/TRK/3417), the purpose of which is to provide a more robust means for controlling track position and the associated gauge clearances and interfaces with other fixed assets.

4.4.1 Measuring the longitudinal level

For a smooth train run and comfortable ride, the longitudinal gradients should be as consistent as possible. This needs to be observed for sighting in particular, as each high point will usually mark a gradient break. This section will describe the non-geodetic procedures for measuring the longitudinal level.

4.4.1.1 Sighting

Sighting is the simplest form of precision measurement for the longitudinal level. Various measuring devices are available for this.

Fig. 4-10 Sighting device (optic)

The sighting device consists of a stand with a clamping device for fastening it to the rail. The sighting instrument (optic) is put on the stand. The spirit level is used to establish a vertical position of the stand. The measuring staff also has a spirit level to ensure vertical positioning.

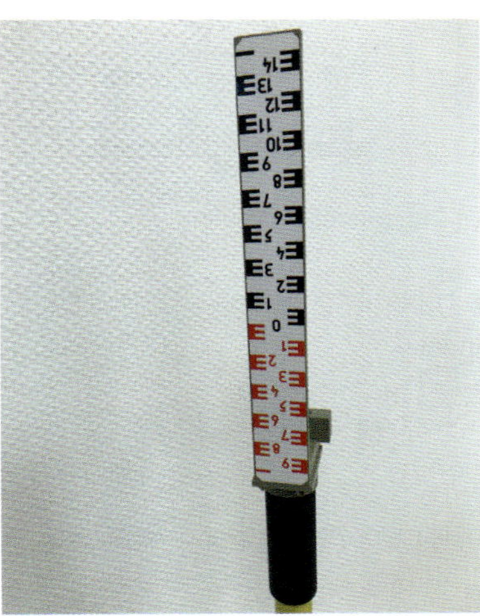

Fig. 4-11: Measuring staff of the sighting device

The measuring scale of the sighting staff has an E-shape graduation. Each E-element is 10 mm high and each horizontal line within the E-element is 2 mm high (see also Fig. 4-14). The zero point of the measuring staff for the sighting device is at the intersection of the red and black marked E-elements.

Fig. 4-12: Switchable sighting and levelling instrument

In addition to the pure sighting device, combined levelling and sighting devices are available. For sighting the measurement is taken at an arbitrary incline. On the other hand, for levelling the measurement must be taken horizontally aligned. This requires an absolute vertical positioning of the stand and fitted measuring optics. For this purpose, the combined sighting and levelling devices have bubble tubes for both the longitudinal and cross direction, or they have a circular bubble. In the example shown, the switch between sighting and levelling is carried out using the switching screw fixed at the top.

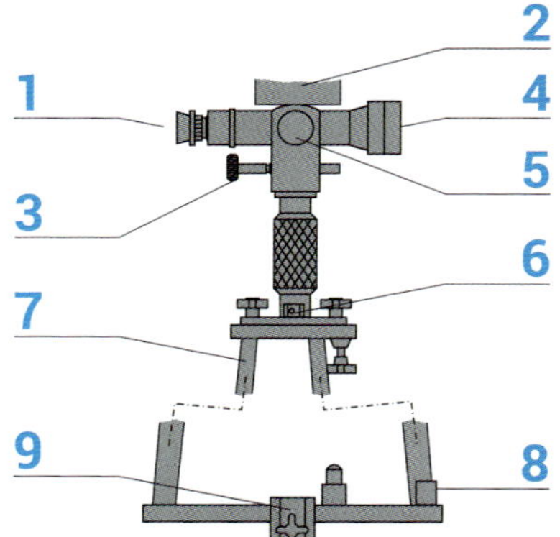

Fig. 4-13: Operating elements of the optical instrument

The sighting device has several operating elements:
1. Ocular with dioptre adjustment
2. Rough sighting setup
3. Tilting screw for adjusting the incline of the telescope
4. Telescope
5. Focusing drive (focusing)
6. Millimetre scale for height adjustment (adjustment of target lifting values at measurement point)
7. Stand
8. Spirit level
9. Clamping device

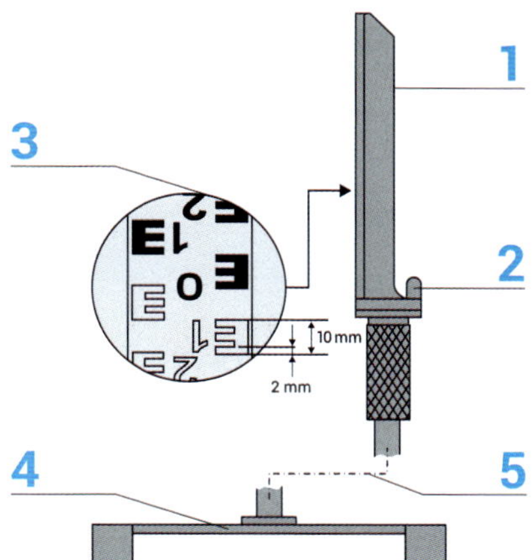

Fig. 4-14: Elements of the measuring staff

The components and scaling of the measuring staff of the sighting device:

1. Scale strip
2. Spirit level
3. Scale
4. Mounting angle
5. Stanchion

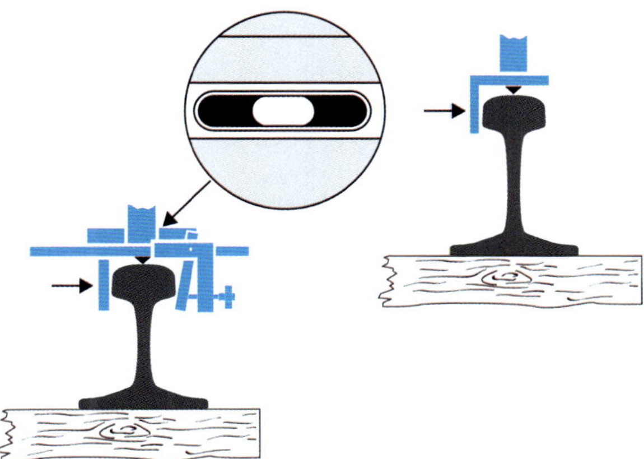

Fig. 4-15: Positioning the measuring staff and stand

When setting up the levelling instrument and the measuring staff on the reference rail, attention needs to be paid that the mounting angles are pressed against the inner edge of the rail. Using the spirit levels, they are put in a vertical position.

For the deployment of track construction machines, **high point – high point sighting** (HP-HP sighting) is used where there is no existing fixed point marking (e.g. when correcting isolated defects). To do this, the high points are determined visually at a distance of 70 m maximum (optical range of sighting device) on the track.

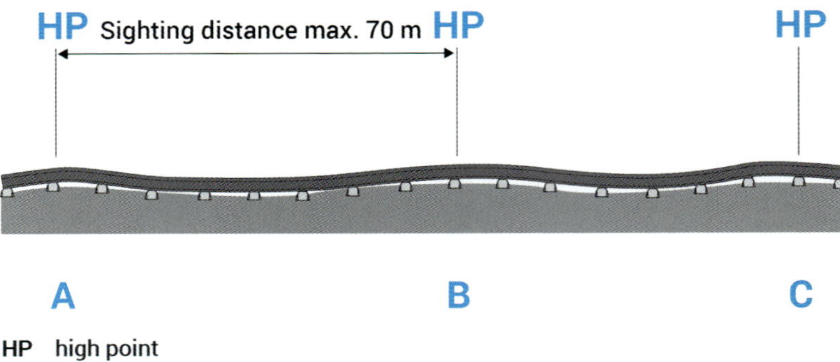

Fig. 4-16: Sighting the visual point distance (high point)

The sighting device is placed two to three sleepers before or after high point A on the low or reference rail ("intersecting" the high point), and the basic lift is adjusted (basic lift = settlement value, usually 10 – 20 mm). To determine the lifting values at the high points, the defect in the cross level at the high point needs to be established. This is done using a cant measuring device (cross level gauge). Then the basic lift is aimed at on the measuring staff on high point B. In the last step, the measuring staff is moved closer by approx. 5 m (corresponds to approx. 7 sleepers), and the lift is written down on the sleepers.

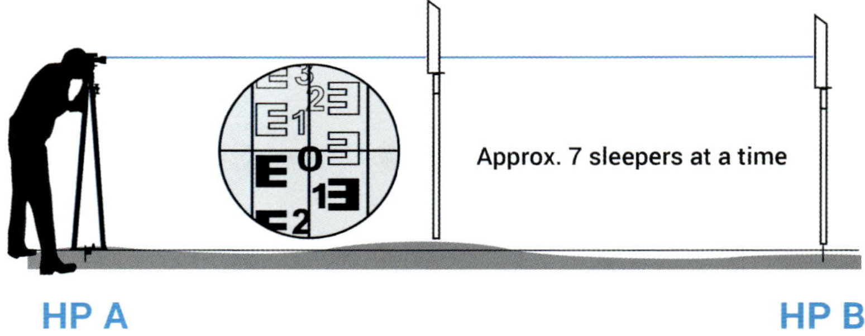

Fig. 4-17: Measuring 7 sleepers at a time in the direction of high point A

This makes it possible to determine the lifting values that correct an irregular track geometry of the longitudinal level between the high points. However, each high point represents a change in the longitudinal gradient. On a route with overhead catenary it may be necessary to ensure that the new track gradient is complementary to the wire gradient.

In order to support the maintenance tamping operation, **start and end ramps** have to be created at the start and end of the section of measurement in order to enable a continuous transition from the existing track to the track being worked on. The ramp gradients depend on the permitted maximum line speed and are defined by the infrastructure managers.

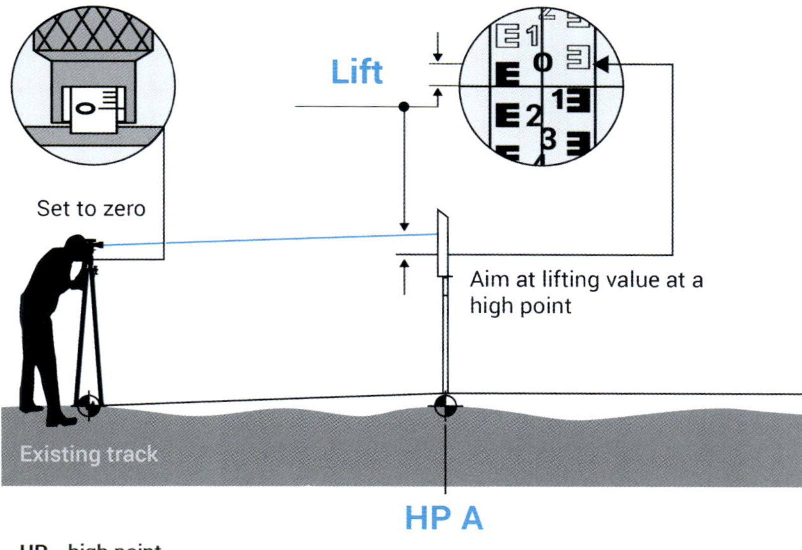

Fig. 4-18: Measuring the start ramp

In order to carry out measurements using the start ramp, a suitable connection point has to be found on the existing track at an appropriate distance (gradient of start ramp has to be observed). Then the height adjustment on the sighting device is set to zero and the desired lifting value is aimed at on the measuring staff at the high point.

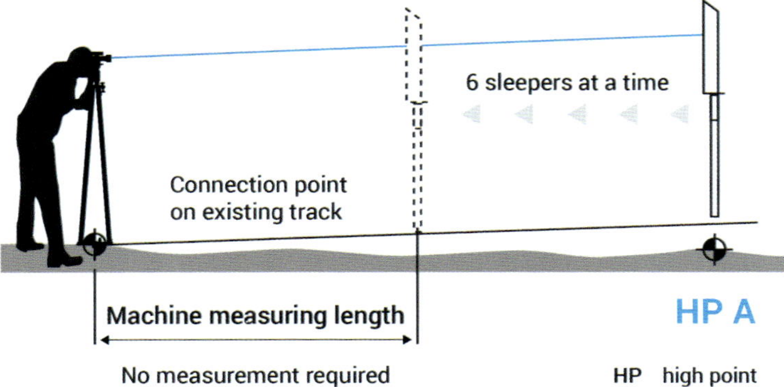

Fig. 4-19: Measuring the start ramp every 6 sleepers

From high point A, 6 sleepers at a time are measured in the direction of the connection point on the existing track. The area of the machine measuring length from the connection point towards the high point does not need to be determined. The settings on the sighting device must not be changed as the machine front extension (to guide the machine at the front transducer) still needs to be measured.

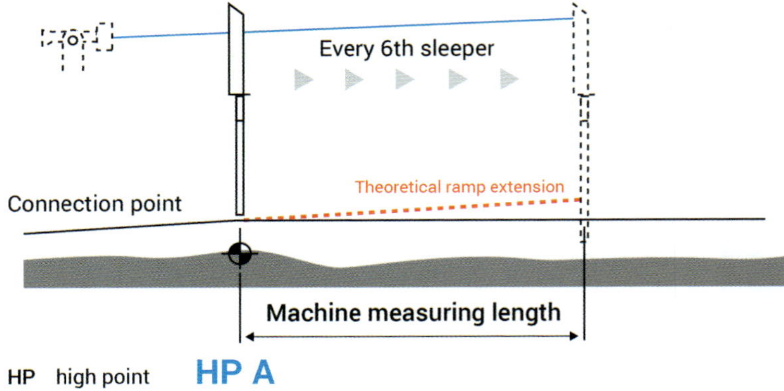

Fig. 4-20: Measuring the machine front extension of the start ramp

Without changing the settings on the sighting device, the machine front extension has to be measured from high point A on every 6th sleeper on a length that corresponds to the machine measuring length.

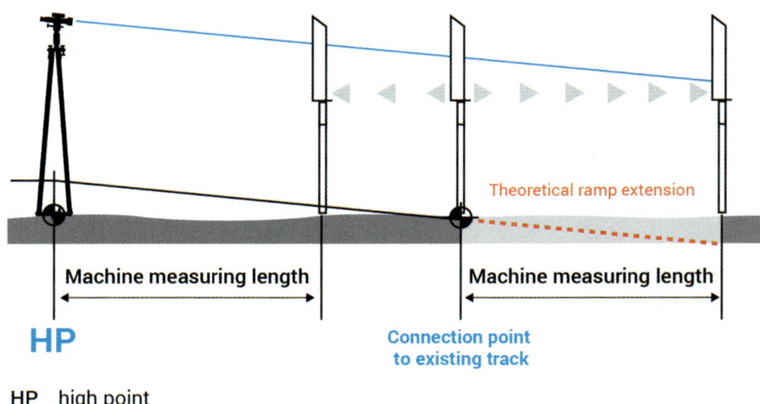

Fig. 4-21: Creating the end ramp with machine front extension

This **machine front extension measurement** enables the front transducer of the machine to be guided and thus for the straight progression of the longitudinal level to be created.

For pre-measuring using the sighting method on a reverse curve with **canted transition ramps** (see section 3.7) the following steps have to be performed:

The cross level has to be checked at both ends of the ramp (using a cross level gauge), also the longitudinal level. To do this, the required settlement value has to be determined. Then the sighting device is set up at the ramp end (RE) of the non-canted rail and the levelling staff at the ramp end of the canted rail.

After that, a zero sighting procedure is performed (taking into consideration the settlement value). The measuring staff can now be moved forward to ramp start/ramp start (RA/RA), and the level can be determined for every 7th sleeper from RA/RA to RE (on the rail that is not canted).

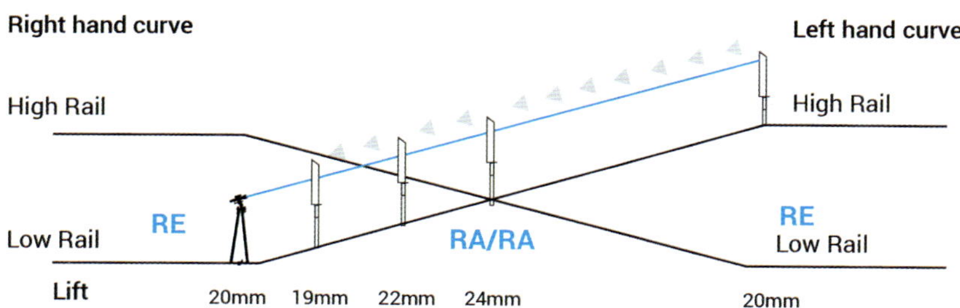

RE   ramp end
RA/RA is the point of contraflexure between the two curves

Fig. 4-22: Sighting the reverse curve with canted transition ramps, part 1

Subsequently, the cross level is checked at RA/RA (using a cross level gauge); any cross level defects and the change of the rail need to be taken into consideration. Finally, the sighting device is repositioned to determine the lift in the remaining part of the reverse curve from RA/RA to RE.

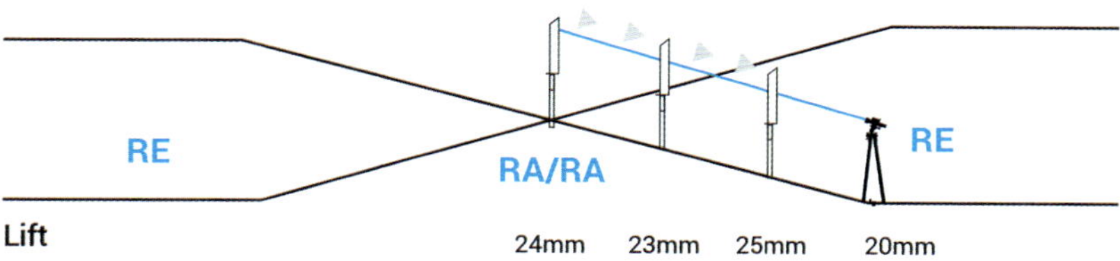

RE   ramp end
RA/RA is the point of contraflexure between the two curves

Fig. 4-23: Sighting the reverse curve with canted transition ramps, part 2

If there is a change of the longitudinal gradient (gradient change) within the section where the track geometry has to be pre-measured by sighting, the following steps will be required to determine the **lifting values at the gradient change using the sighting device**:
1. Determine the gradient change point (from the longitudinal section; if not available, determine visually).
2. Determine the lifting values at the start and end of the curve according to the marking values or the settlement value.
3. Determine the radius in line with the locally permitted speeds and convert tangent line lengths into sleeper bays.
4. Perform the sighting from the start to the end of the curve and determine the actual longitudinal level.
5. Enter the actual values in the calculation table.
6. Enter the target values from the calculation table.
7. Determine the lift from the difference between the target and actual values.

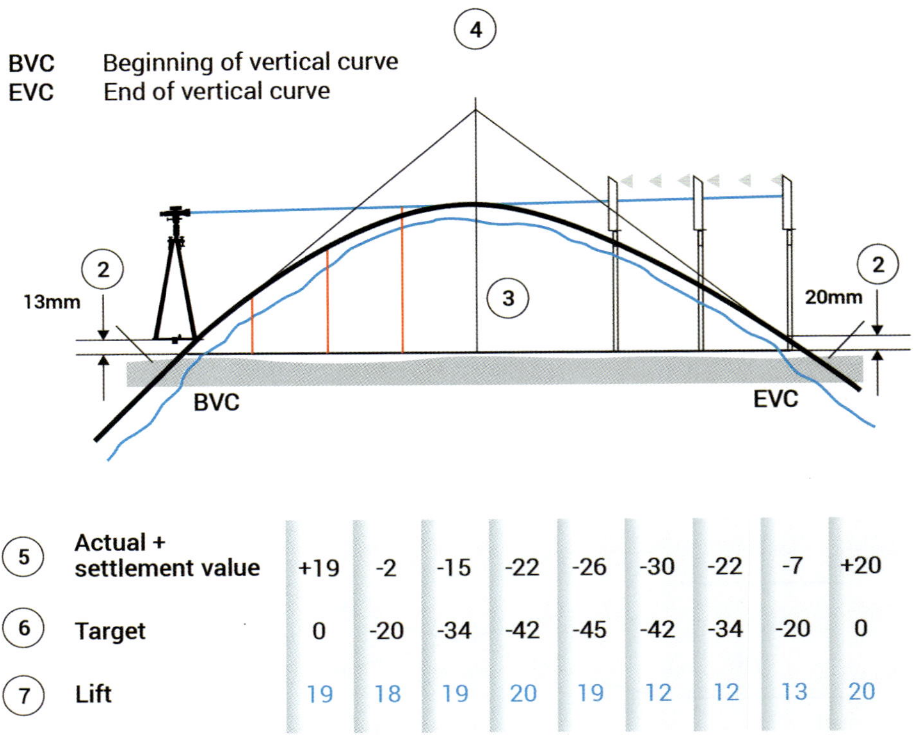

5	Actual + settlement value	+19	-2	-15	-22	-26	-30	-22	-7	+20
6	Target	0	-20	-34	-42	-45	-42	-34	-20	0
7	Lift	19	18	19	20	19	12	12	13	20

Fig. 4-24: Sighting in a gradient change

The example of a gradient change shows a hump (the centre of the curve is below the track). Proceed accordingly for dips (the centre of the curve is above the track).

The large image in Fig. 4-25 illustrates the procedure for **sighting the longitudinal level in measured track**.

If the track to be measured is marked at fixed points (e.g. masts) and there is no gradient change within the measured section (**longitudinal level measurement with constant gradient and marking** at the fixed point), the pre-measurement can be performed as follows (see Fig. 4-25 below left):
1. The lift at the mast bolts or spigots is determined taking into consideration the actual dimensions from the longitudinal section.
2. The determined lifting value is set on the sighting device at mast 1.
3. The determined lifting value is aimed at, at mast 2.
4. Sighting is performed, and the lifting values are read every 7th sleeper.

The **longitudinal level measurement with marking with known target measurements** can be used if the target track geometry is marked at fixed points (e.g. masts) and the required target measurements have been calculated for the respective measurement points depending on the target track geometry (e.g. ÖBB).

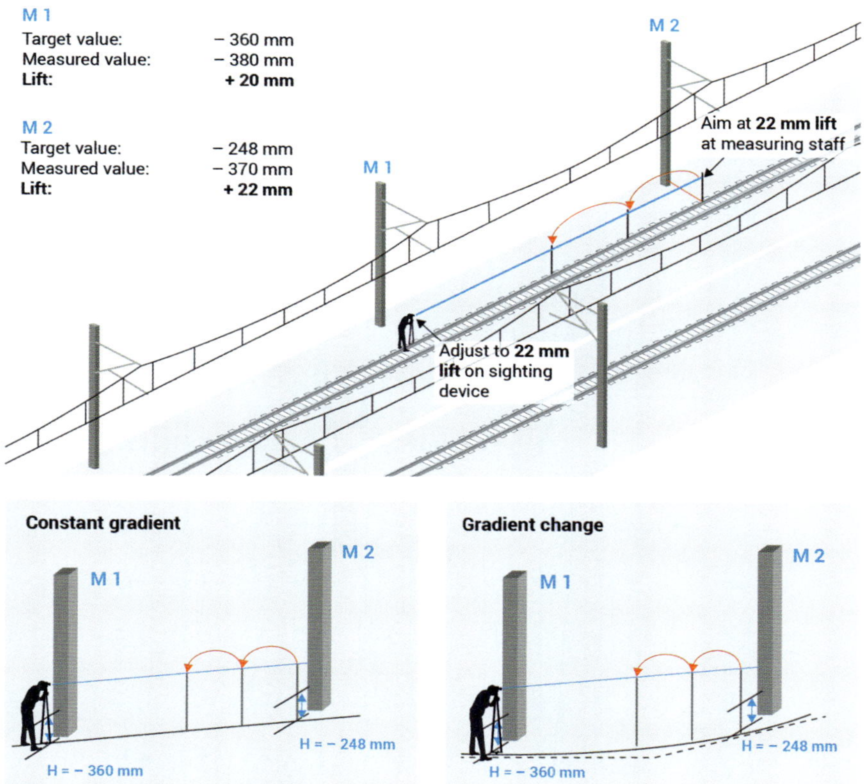

Fig. 4-25: Longitudinal level measurement with fixed point marking (studs in the masts)

For the longitudinal chord procedure, a chord (optical or laser) is tensioned between two marking points. At DB this is based on the respective target values from the route layout plans (Fig. 4-26) or turnout level plans (Fig. 4-27); at ÖBB it is based on the marking lists. These target values are compared with the actual values measured on site to obtain the correction values for level and alignment. The correction values need to be determined for the long chord points (every 5 m), see Fig. 4-25 bottom right.

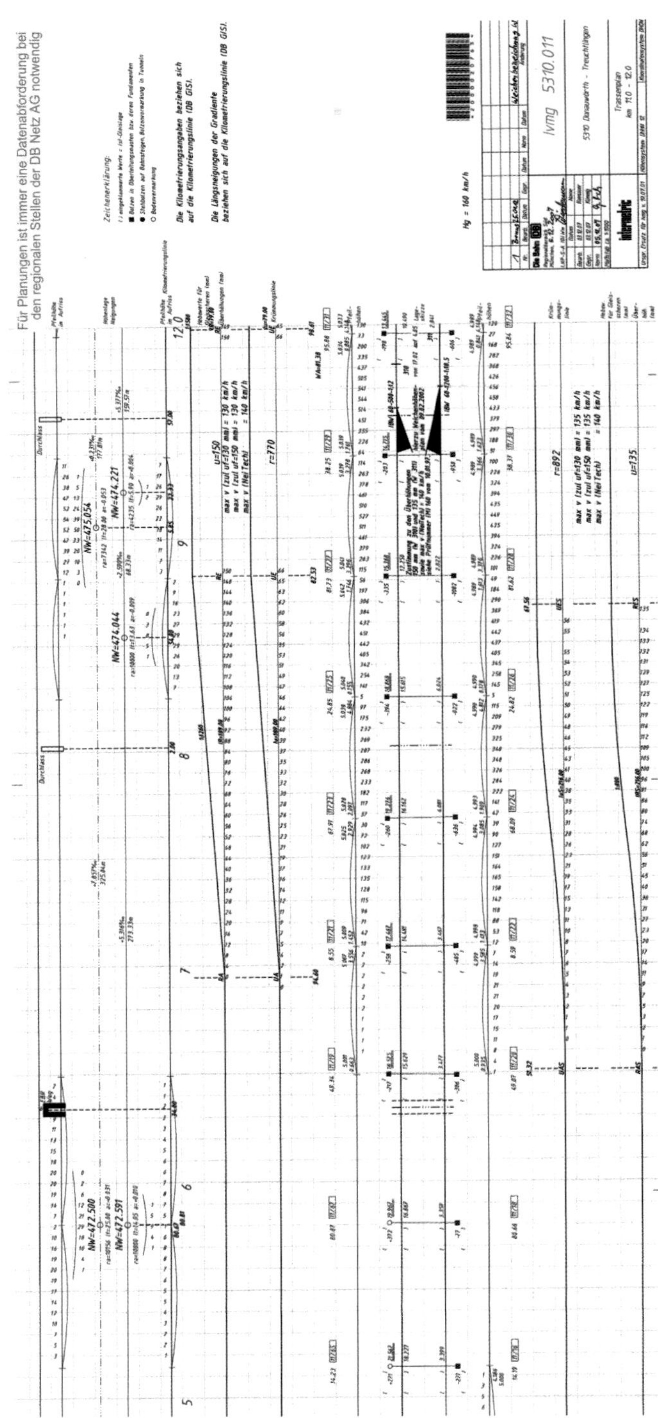

Fig. 4-26: Route layout plan of DB Netz AG

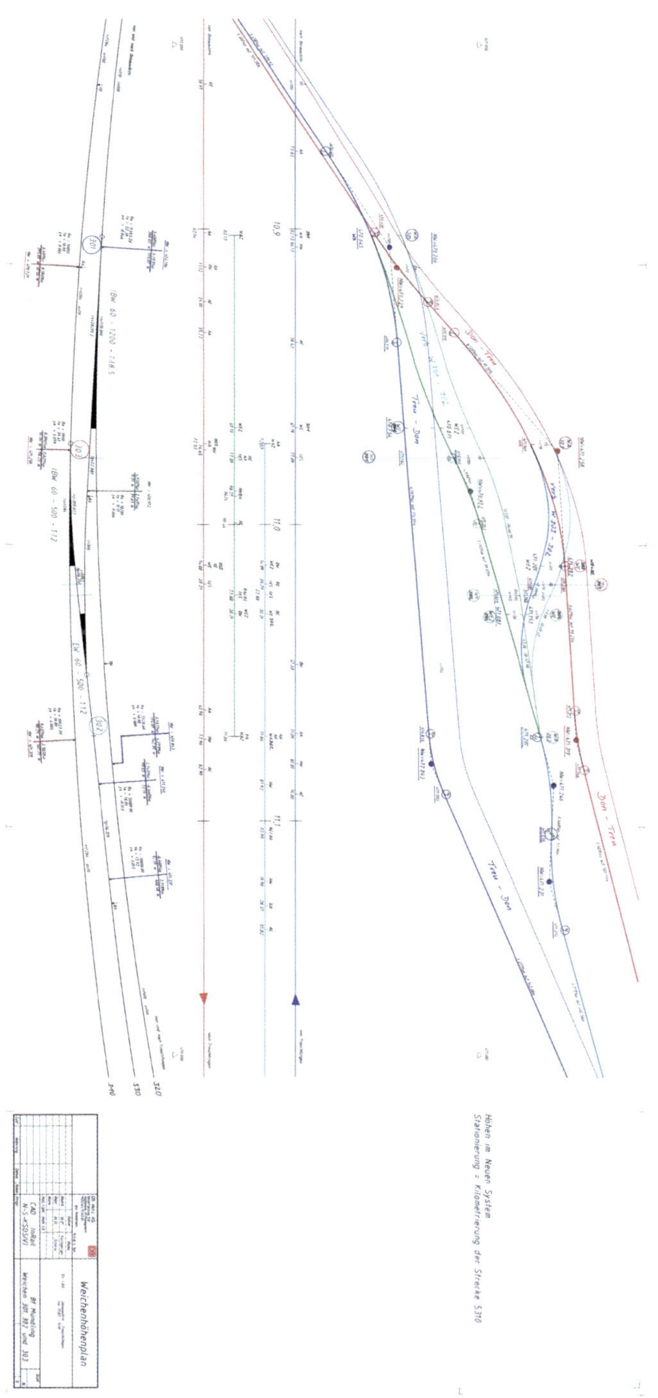

Fig. 4-27: Turnout level plan of DB Netz AG

4.4.1.2 Levelling

In contrast to sighting, levelling is defined as measuring level differences in a horizontal plane. This is carried out using a levelling device. It consists essentially of a tripod and a telescope with crosshairs. The target axis of the telescope is levelled horizontally using spirit levels.

During levelling, the values for the horizontal target line are read at vertically positioned measuring staffs (levelling staffs). The values can be read manually; with digital levelling devices, the values are read and stored automatically.

The level difference between two points is determined from the associated measurements, called backward (B) and forward (F) measurements. It is calculated from the difference between the backward and forward reading (rise = +; fall = −).

The so-called levelling equation is:

$$\Delta h = R - V \qquad (4\text{-}1)$$

Δh: level difference

$$\Delta h = R - V = 303 - 121 = +182 \text{ mm}$$

Example: level of start point

$$A = 85.604 \text{ m NHN}$$

The level of the end point B is obtained by adding up the level difference:

$$H_B = H_A + \Delta h = 85604 + 182 = 85786 \text{ mm} = 85.786 \text{ m NHN}$$

Fig. 4-28: Principle of levelling

If start point A and end point B are so far apart that they cannot be covered by a single measurement, the section of measurement has to be extended by transposing it. Starting from point A, the instrument is repositioned for determining the levels, which are lined up, until end point B is reached.

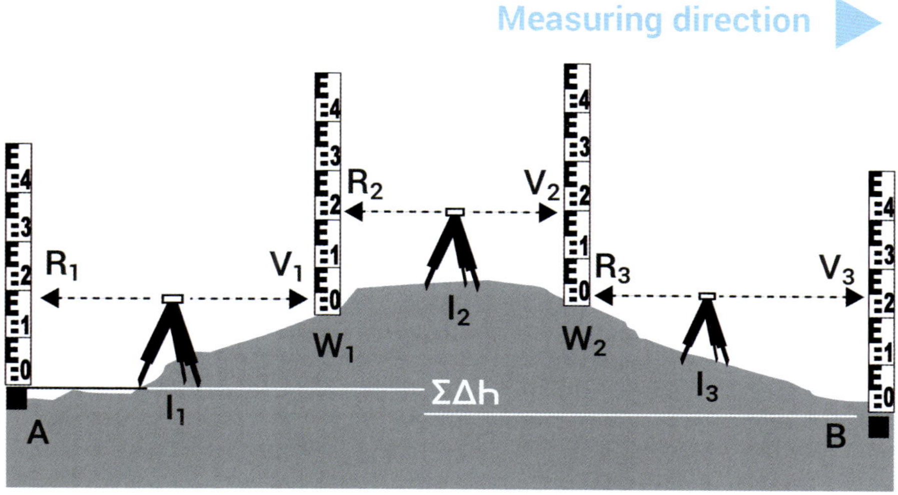

Fig. 4-29: Repositioning the instrument during levelling

Method of repositioning:

1. Positioning device in l_1; reading at R_1 and V_1.
2. Repositioning device to l_2; reading at R_2 and V_2.
   The height of the levelling staff mounting points at the change points W_1 and W_2 must not be changed between the forward and backward measurements.
3. After repositioning the device to l_3 with backward measurement R_3 and forward measurement V_3, the target has been reached.
   The overall level difference Δh between point A and point B is calculated from the sum of all individual level differences.

Levelling should always be continued to a second fixed point with a known level and completed there. If starting at a fixed point with a known level and ending at another fixed point with a known level, it will be possible to check the sums of the individual measurements and the known level differences of the fixed points ($H_B - H_A = \Delta h_{Targ}$).

If there is no second fixed point nearby, the levelling process can be taken back to the first fixed point (levelling loop). Measurement check: $\Delta h_{Targ} = 0$. However, with this type of checking, any changes at the connection point are not recognised.

This type of measurement is primarily used for surveying, but can also be used for pre-measuring (with target-actual comparison) and for pegging out the levels. Both local networks (e.g. a fixed point as a starting level) and national level systems, national coordinate systems or railway-specific surveying points can be used as reference levels.

4.4.2 Measuring the alignment

This section will describe the non-geodetic procedures for measuring the alignment.

4.4.2.1 Moving chord or string chord measurement

The simplest way of measuring alignment is the centre moving chord measurement. This determines versines that can be used to depict an approximate alignment of the track curvature. With the string or moving chord procedure, a chord with a precisely defined length is tensioned, and the centre versine in the curve is measured (to mm accuracy). Chord length s is 10 m with R < 300 m and 20 m with R ≥ 300 m.

For the manual measurement method, the measurement points have to be set out first. With a chord length of 10 m, the measurement points have to be distributed at 5 m distance intervals (accordingly, the measurement point distance is 10 m with chord lengths of 20 m). Then the versine (distance from the chord to the running edge at the measurement point) is determined for each measurement point; to do this, the chord ends are put against the running edges of the measurement points that are all 5 m apart. Once measuring has been completed at a measurement point, the process is repeated at the next measurement point (at a 5 m distance in the example).

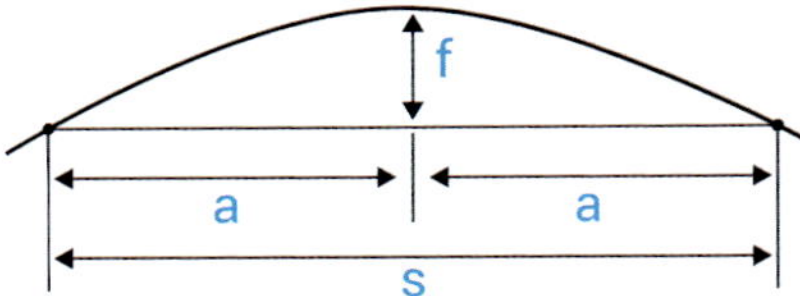

Fig. 4-30: Centre versine measurement

R in [m]:   radius in curve

s in [m]:   chord length

f in [m]:   versine

This versine measurement is used to determine the radius or the versine of the curve:

$$R = \frac{125 \cdot s^2}{f} \qquad (4\text{-}2)$$

$$f = \frac{125 \cdot s^2}{R} \qquad (4\text{-}3)$$

With the method of the off-centre versine measurement, the radius (or target versine) of the curve can be calculated at each point within the chord. Due to the variation in chord length, this calculation and measurement method is also used for long chord measurement.

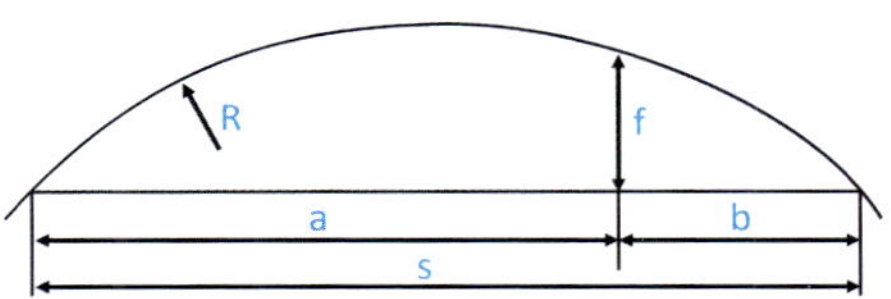

Fig. 4-31: Off-centre versine measurement

f in [mm]:   versine

s in [m]:   chord length

a in [m]:   distance from start point to measurement point

b in [m]:   distance from measurement point to end point

R in [m]:   radius

Calculation formulas for off-centre versine measurements in the curve:

$$R = \frac{a \cdot b \cdot 1000}{2 \cdot f} \tag{4-4}$$

$$f = \frac{a \cdot b \cdot 1000}{2 \cdot R} \tag{4-5}$$

Determining the versine at any point in the transition curve:

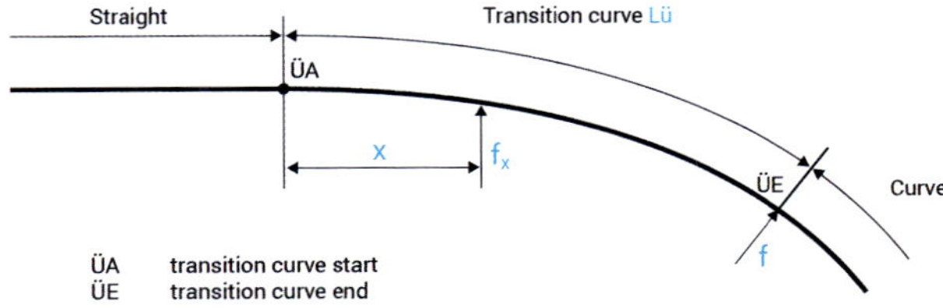

Fig. 4-32: Determining the versine in the transition curve

ÜA: transition curve start

ÜE: transition curve end

x in [m]:   distance to transition curve start

f in [m]:   versine in the curve

f_x in [m]:   versine at measurement point in transition curve

Lü in [m]:   length of transition curve

Determining the versine in the transition curve:

$$f_x = \left(\frac{f}{L\ddot{u}} \right) \cdot x \tag{4-6}$$

Another method for determining the versine at any point in the transition curve:

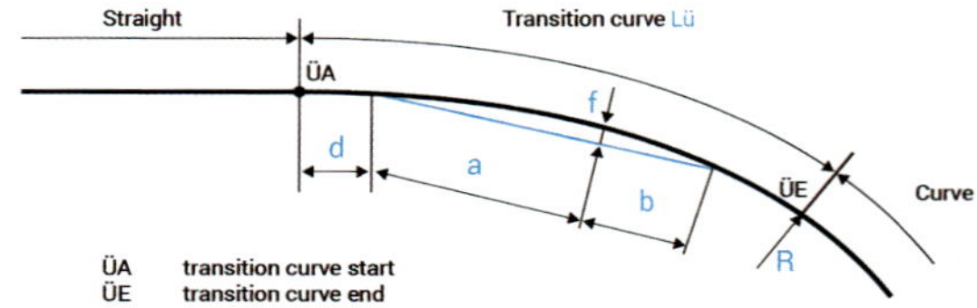

Fig. 4-33: Versines in the transition curve

f in [mm]:   versine
ÜA: transition curve start
ÜE: transition curve end
d in [m]:   distance from transition curve start to first chord start point
a in [m]:   distance from chord start point to measurement point
b in [m]:   distance from measurement point to chord end point
R in [m]:   radius
Lü in [m]:   length of transition curve

$$f = \frac{a \cdot b \cdot 1000}{6 \cdot L\ddot{u} \cdot R} \cdot (3 \cdot d + 2 \cdot a + b) \tag{4-7}$$

Calculating the radius at any point in the transition curve:

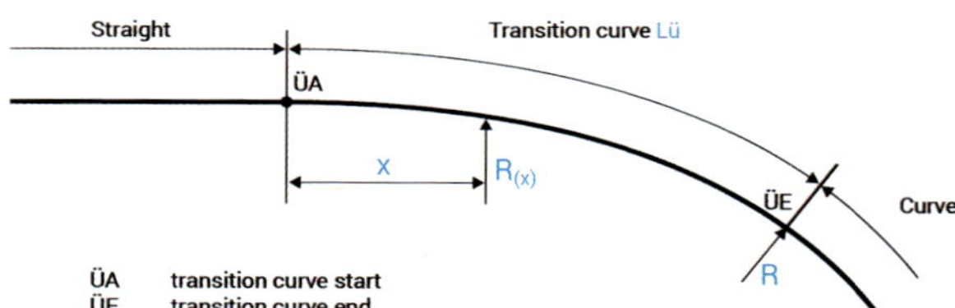

Fig. 4-34: Determining the radius at a defined point in the transition curve

ÜA: transition curve start
ÜE: transition curve end
x in [m]:   distance to transition curve start
a in [m]:   1/2 chord length
R in [m]:   curve radius
$R_{(x)}$ in [m]:   radius in transition curve
Lü in [m]:   length of transition curve

Radius at any point in the transition curve:

$$R_{(x)} = \frac{R \cdot L\ddot{u}}{x} \tag{4-8}$$

4.4.2.2 Long chord measurement

Long chord measurement can be used for surveys (to determine the target track geometry) and pre-measuring (to determine the slueing values). A general precondition will be a geodetic survey of the measurement points at the masts and measurement of the track geometry at the respective end points of the chord (at each mast).

There are different types of measurements in the DACH countries. At DB and SBB the measurement of the alignment refers to the track centreline, while at ÖBB the reference point will usually be the running edge of the outer rail of the curve.

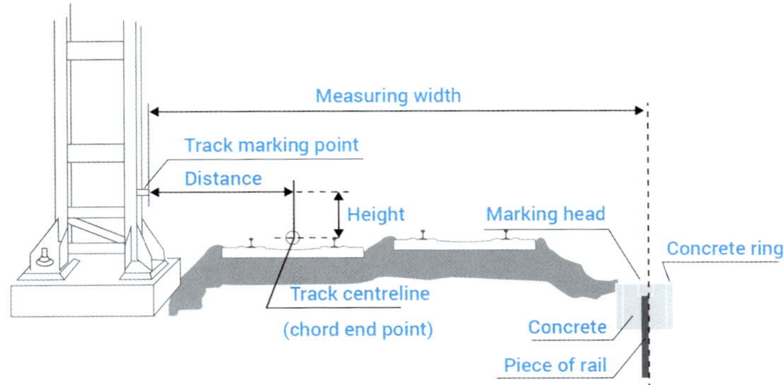

Fig. 4-35: Marking at masts (example: DB)

The principle of long chord measurement is as follows:

If there is an overhead line present, fixed marking points are set up in line with the catenary support distances (on both sides of the track at DB). If there is no overhead line, the marking is performed at fixed points made out of concrete. A point of intersection is determined between the connecting line of opposite marking points and the track centreline. At approx. 5 m distance intervals, target versines are defined on the long chord (polygon chord) that reflect the precise curvature of the track. [17]

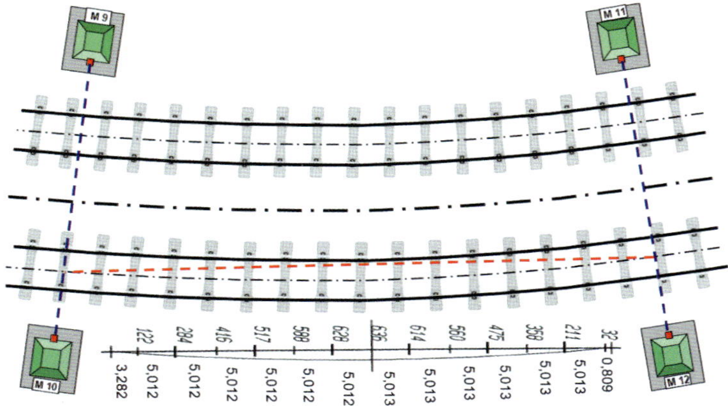

Fig. 4-36: Measured long chord procedure

The long chord measurement procedure is used to determine the lifting and slueing values for pre-measuring. The target distance from the track centreline to the marking point has to be determined for each marking point, resulting in the target position of the outer rail in the curve.

Tensioning a chord (optical or laser) from mast to mast will provide measurements for the intermediate points every 5 m; these are compared with the target values. The difference between the actual and target values is the slueing value or slues that the tamping machine needs to work with. This measurement procedure is also referred to as the offsetting of versines.

As of 01.01.2020, DB has replaced the long chord measurement with geodetic pre-measuring to be integrated into the DB reference network "DB_REF"; long chord measurement will only be permitted in exceptional circumstances.

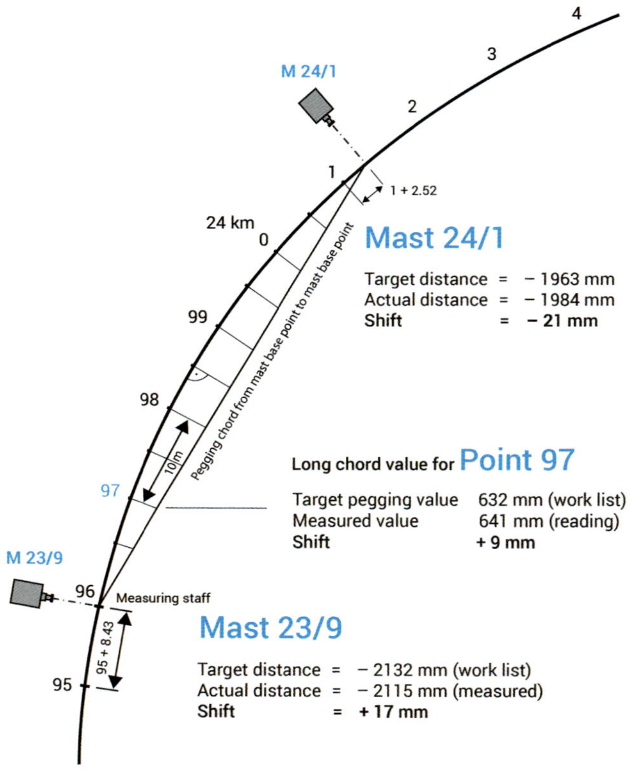

Fig. 4-37: Long chord measurement for pre-measuring

4.4.3 Systems for measuring longitudinal level and alignment

The measurement procedures dealt with so far are manual measurement procedures. Different measuring systems are used for measuring the longitudinal level and the alignment. Therefore, a separate work step is required for each measurement. These procedures have been developed further and partly automated to improve and optimise the work processes and quality, leading to higher efficiency.

4.4.3.1 EM-SAT track recording car

EM-SAT is an automated measuring system using the principle of long chord measurement for measuring level and alignment.

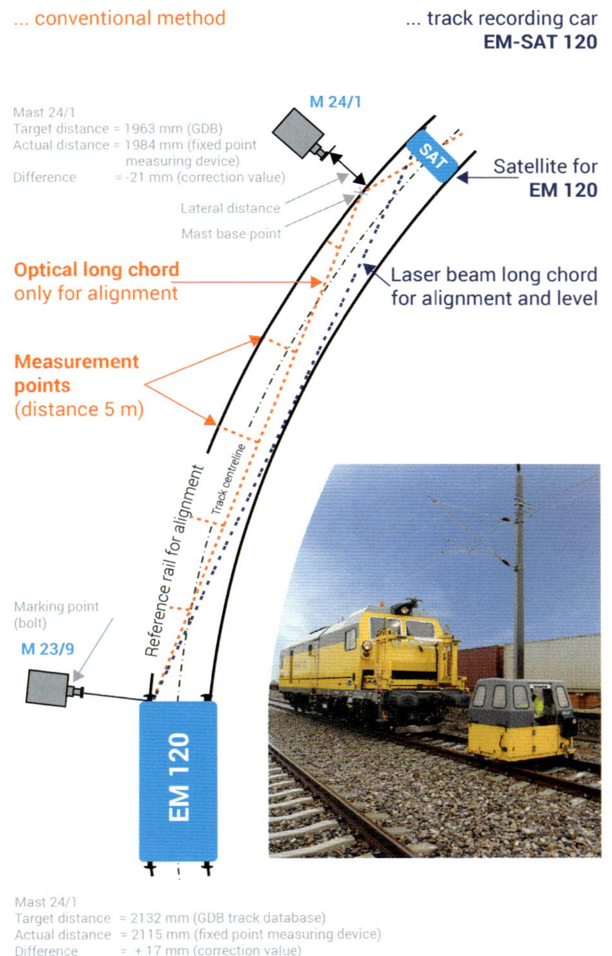

Fig. 4-38: EM-SAT long chord measurement principle

The machine is self-propelled and has an offset, electrically powered measuring satellite vehicle to increase the measuring base. The measuring chord is tensioned from one mast to the next via a laser between the main machine and the satellite. The fixed points on the mast are measured with fixed point measuring devices (integrated into the main machine of newer models). While the satellite vehicle retains its position, the main machine with the laser receiver is continuously moving towards the satellite vehicle and at the same time measures the position changes in height, alignment and cant.

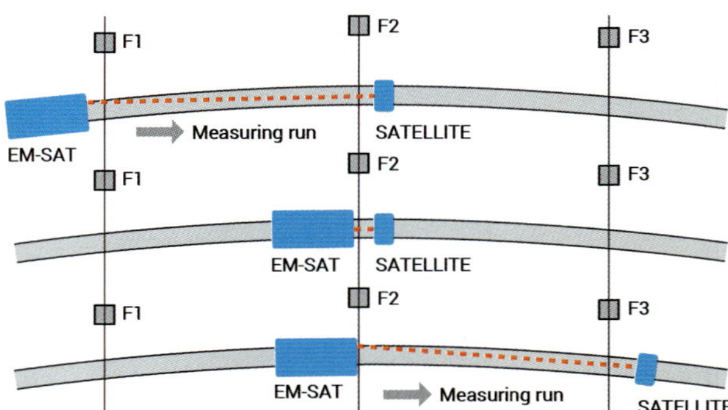

Fig. 4-39: EM-SAT measuring process

As the track geometry data had been input into the on-board computer before the start of measuring, the lifting and slueing values can be output as soon as a section of measurement from one mast to the next has been completed. In addition, the measurement data is stored in the vehicle's on-board computer and after completion of the measuring process can be directly transmitted to the tamping machine or for further processing (change of route layout, approval, etc.).

This method of measurement is called a cyclical measurement as the measurement is carried out from one mast field to the next. Since the measuring systems are integrated into a machine, the measurement is carried out on the loaded track.

The EM-SAT can be additionally equipped with other measuring systems, such as:

– Ballast profile measuring system: This is used for determining the currently available track ballast volume taking into consideration the lifting values and thus calculating additional requirements as well as existing surpluses of track ballast. The measurement outputs form the basis for optimised ballast management.
– Systems for measuring constraints such as platforms, ramps, other structures, level crossings, and also for measuring the position of the catenary.
– GPS systems combined with reference procedures (e.g. differential GPS (DGPS))

The EM-SAT can perform surveys, pre-measuring and post-measuring runs.

Two EM-SAT machines were deployed in the UK on the West Coast Main Line between 2003 and 2010.

Fig. 4-40: EM-SAT: the satellite in the front with the main machine behind

4.4.3.2 Combined levelling and lining laser

The combined levelling and lining laser is used in combination with a tamping machine. This is a measuring device with a laser that is positioned on the straight track in front of the tamping machine. The laser points to the receiver unit on the tamper. During the measuring run, the machine moves towards the combined levelling and lining laser and continuously measures the track geometry. The measurement data is recorded in the on-board computer, and the correction values can be changed before maintenance work.

Fig. 4-41: Combined levelling and lining laser

4.4.3.3 Curve laser

The curve laser is a development of the combined levelling and lining laser. This is also a measuring system used in combination with a tamping machine.

Fig. 4-42: Curve laser

With the track geometry known in advance, the curve laser is set up at a point with a known position and level at an appropriate distance from the tamping machine. Any lifting and slueing values at the setup point are recorded in the on-board computer, as is the target geometry. The on-board computer can calculate the target geometry as specified and during the measuring run can deduce the difference between the target and actual geometry, also in curves and transition curves. Curve laser has been used successfully on Network Rail's Western Route.

4.4.3.4 PALAS system

The PALAS measuring system uses a geodetically created measuring system as a reference. Measurement points at catenary supports serve as fixed points. This system is primarily in use at SBB.

Fig. 4-43: PALAS measuring system

The PALAS system consists of a laser with an integrated receiver unit attached to the tamping machine and triple reflectors attached to the fixed points. The laser transmits the laser beam, which is reflected by the triple reflectors and picked up again by the receiver unit. Based on the measurement of the fixed points, the position of the machine (at the front tensioning trolley) can be calculated and compared with the geodetic target position. From this, the lifting and slueing values for achieving the target position are derived. A precondition is that the view of the laser of one or more fixed points must be ensured at all times.

Since the measurement is carried out at the same time as tamping, there is no need for an additional track closure. However, it must be ensured that the survey is free of inconsistencies throughout and that there is always a clear view of the fixed points. If the target track geometry cannot be established for various reasons (e.g. lifts or slues are too large, changes to the fixed points, etc.), it will often not be possible to respond in time.

Due to a lack of information on the lifts to be expected during the tamping process, it will be challenging to pre-deposit the right amount of ballast required to establish the appropriate ballast bed cross-section.

PALAS was deployed in the UK on the West Coast Main Line Northern sections between 2003 and 2010.

Fig. 4-44: PALAS system laser scanner

4.4.3.5 Hand-guided measuring systems

Hand-guided measuring systems consist in principle of one or two hand-guided measuring units (measuring trolleys). These systems can generally be divided into three different categories:
- pre-measuring with a two-trolley system using the tachymetric long chord measurement procedure
- geodetic pre-measuring with tachymeter and one measuring trolley
- inertial measuring system on a hand-guided measuring trolley

Pre-measuring with a two-trolley system using the tachymetric long chord measurement procedure uses a tachymeter installed on a track measuring trolley and a measuring prism installed on a second trolley. For positioning itself, the tachymeter measures the distances of two adjacent mast bolts to the track.

As a first step, the tachymeter trolley is positioned vertically to the first reference point on the track. The track marking point is signalled by a prism or measured reflectorless to determine the actual distance to the track centreline and the gradient. As the target values will typically already be known and stored, the deviations of the position, level and cant are displayed directly on site. Then the tachymeter trolley is pushed to the next fixed point, which shall be used as the end point of the chord definition. Here, too, the trolley is positioned vertically to the marking, and a target-actual comparative measurement is carried out. Thus, the required slue and lift against the target geometry is known at both chord end points.

Fig. 4-45: Example of a hand-guided two-trolley measuring system

At the position of the first fixed point measurement (vertically to the first reference point) the prism trolley is now positioned on the track, and the tachymeter, which is still positioned on the trolley at the other chord end point, is turned towards the prism trolley. The subsequent measurement of the horizontal direction, the vertical angle and the incline defines the chord between the two reference points. Following the definition of the chord, the tachymetric measurement value can be compared with the chord definition at every position between the chord end points by moving the prism trolley, and the versine can be calculated from that. The slue and lift for the line points within the chord will be derived by comparing the measured versine with the target versine from the route layout.

For continuous recording of the track geometry, the prism trolley is simply pushed from one chord end point to the tachymeter trolley. Thanks to automatic target tracking and continuous measurement, the current position and the correction values against the target geometry are continuously displayed. The deviations are stored automatically in the user-defined grid. In addition, track-specific points are indicated in order to mark these for synchronisation of the tamping machine in the track. When the prism trolley reaches the position of the tachymeter trolley, the latter is pushed to the next fixed point and the prism trolley takes the previous position of the tachymeter trolley. Following the measurement of the next target-actual comparison, the chord is defined again, and the prism trolley is pushed along the track up to the tachymeter for measuring within the chord.

Thanks to the tachymetric measurement at the fixed points, the chord measurement is integrated into the absolute reference system. For each measurement point in the chord, absolute three-dimensional coordinates for the position and level of the left and right rail as well as the track centreline can be output. This procedure corresponds broadly to the conventional long chord procedure, with level, alignment and cant measurements being carried out in one work step.

Fig. 4-46: Another example of a hand-guided two-trolley measuring system

For **geodetic pre-measuring with tachymeter and one measuring trolley**, the tachymeter is positioned in a free station setup to several mast bolts with known coordinates. Then the measuring prism on the track measuring trolley is calibrated, and the hand-guided measuring trolley is moved along the track. The self-tracking tachymeter continuously measures the position of the hand-guided measuring trolley. Here, too, the track slue and the required lifting values are calculated from the measured position and by computing the difference to the target track with known coordinates.

The advantage of a geodetic survey is primarily in the result of the free station setup. By using several connection points for determining the location, the quality of the fixed points can be checked at the same time and any unstable markings identified.

Fig. 4-47: Hand-guided measuring system with tachymeter and a trolley (example: LEICA GRP3000)

For pre-measuring using an **inertial measuring system on a hand-guided trolley**, the relative track geometry is determined using an inertial measurement unit (IMU).

Fig. 4-48: Example of an inertial measuring system: Trimble GEDO IMS [18]

The IMU is a multi-sensor system that can capture the free movement of a body in space. The movement of the body is captured along the respective coordinate axis using three vertically stacked acceleration sensors.

Three gyros capture the rotation of the body around the x-, y- and z-axes so that a three-dimensional trajectory (track) can be derived from combining this measurement data. If the starting position of the body in any coordinate system is known, the coordinates in the reference system will be available for each point of the measured trajectory. Through measuring synchronisation points recorded on coordinates, the effects from the drift (deviations of the gyros, mainly at a standstill) can be incorporated.

For recording the track geometry, the IMU is fitted on a track measuring trolley. The track measuring trolley is additionally equipped with a gradient sensor, a gauge sensor and a distance measuring wheel. This means that the relevant parameters such as cant, track gauge and relative track geometry are recorded in one work step.

To include the fixed points (masts, track marking points, track checkpoints), the systems have laser measuring devices (horizontal and vertical distance measurement). Including the marking points will subsequently yield a lifting and slueing value to the target position for each point of the trajectory.

The lifting and slueing values can be transmitted directly to the machine or the infrastructure manager's office.

Fig. 4-49: Hand-guided trolley to measure cyclic top track faults (Example shown is the Abtus ABT 7000) [19] GBW

4.4.3.6 The Hallade method of curve realignment

The Hallade or moving chord method of curve measurement and realignment had been the classic means of measuring and correcting horizontal curve geometry in the UK for many years. It was pioneered by the London Midland and Scottish Railway in the 1930's when they introduced the method to their track maintenance staff based on a chord length of 96 feet and six inches (29.413 metres). The process has three phases. The first is the survey, which is carried out by a string line survey of the curve, using overlapping chords touching the head of the rail, measuring and recording the versines. These are then taken to the office, where by either graphical means or the use of a computer programme, the versines that relate to each half chord are adjusted and regularised enabling track slues (positive or negative) at each half chord to be calculated. Mathematically, the total angle turned through by the new curve must be the same as the existing. As versines are a function of the angle turned through by each chord, then, for the design to be complete, the totals of the existing and designed versines must be the same. The computation demonstrates this result when the algebraic sum of the versine differences is zero. The third and final phase is either the setting out of the new curve in the cess, with wooden pegs at each half chord, using nails set to the required track slues, or preparing a slue file for uploading into a tamping and lining machine. Modern traffic patterns and staff safety, together with the development of geodetic on track surveying have now reduced the use of this method of curve surveying. [16 and 20]

Network Rail has a standard for surveying, NR/L2/TRK/3100 [16] in which it describes the processes to be followed in undertaking track surveys for new work and also for track alignment validation surveys. The UK adopts similar techniques with both static and moving geodetic surveys to those in the DACH countries. Since 2013 there has also been considerable development in the UK with the surveying of track and railway infrastructure using demountable train-borne equipment that combines IMU, 360° scan, downward facing scan and HD video capture, all of which are synchronised. This produces a 3D string of geospatial coordinates that can be input to a track design software package to create new track alignments for track renewals, or, if the track is surveyed after a renewal, the data can be used in the assurance and quality process.

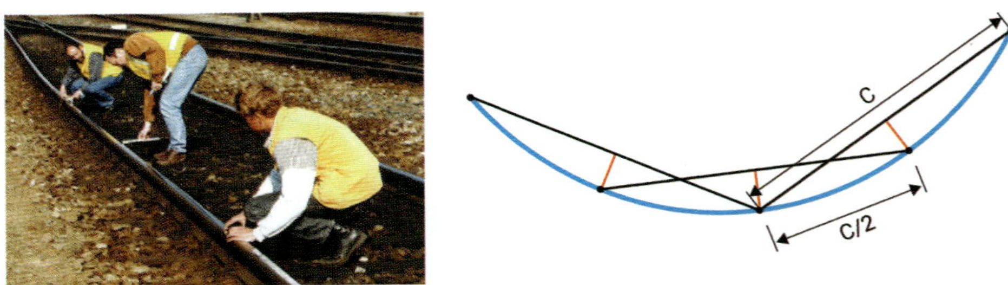

Fig. 4-50: Hallade surveying principles: string line surveying and overlapping half chord versine measurements

4.5 Compensation method

If the precision method for establishing a definite geometrically determined track position, as described in the previous section, is not available, the compensation method can be used to improve the track geometry. Compensation is performed in line with the best-fit method in the progression of the curve or the versine (elevation view and plan view). In contrast to the precision method, the possible track geometry is not defined precisely with this method.

4.5.1 4-point method: compensation method without known route parameters

If the track geometry is not known, the 4-point compensation method will generally be used. For this purpose, the machine has four measuring axles, with the outer two (A and D) used as tensioning trolleys and the two middle ones (B and C) fitted with transducers.

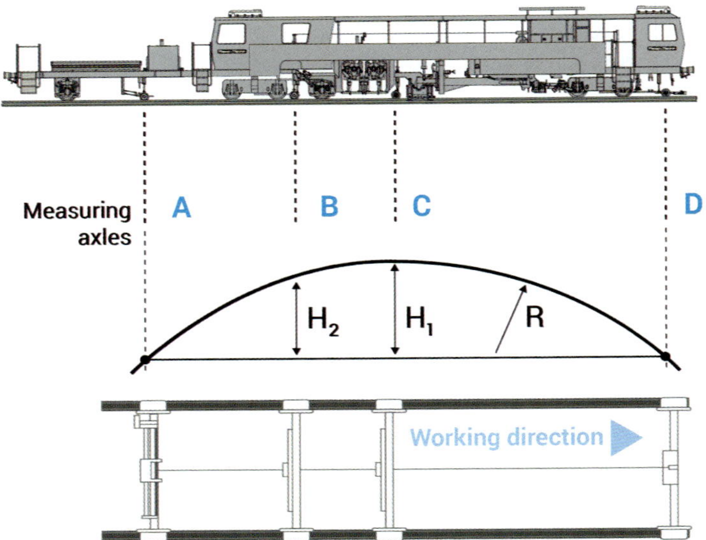

Fig. 4-51: Diagram of 4-point measuring equipment

This method works to the following principle:

The target versine at D of chord section AD is calculated based on the versine at B of chord section AB and using an intercept theorem calculation. (The position of chord AB, which is on the already tamped and lined track, is copied for the front transducer trolley). By shifting measurement point C (at the tamping and lining unit), the lining process is carried out until the chord ratios align.

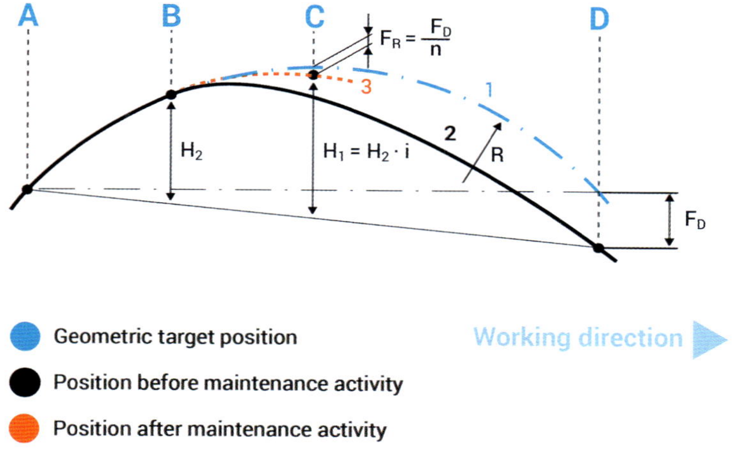

Fig. 4-52: Reducing a defect with the use of 4-point compensation

A residual defect (F_R) remains at measurement point C, depending on the measurement point distances. In the course of the lining work, the measuring axles A and B arrive at the track areas with residual defects and so determine a new basis for the subsequent measuring and lining process. This will result in further deviations from the geometric target position.

By using this method, the track geometry defect can be reduced to about a quarter of its original size.

4.5.2 3-point method: compensation method with known route parameters

For this method, the target track geometry has to be known. The track is scanned at three points (A, C and D), the measuring axle B is omitted. The lining versine at point C is specified from the target track geometry. The lining process is carried out until the target versine H1 is achieved at C.

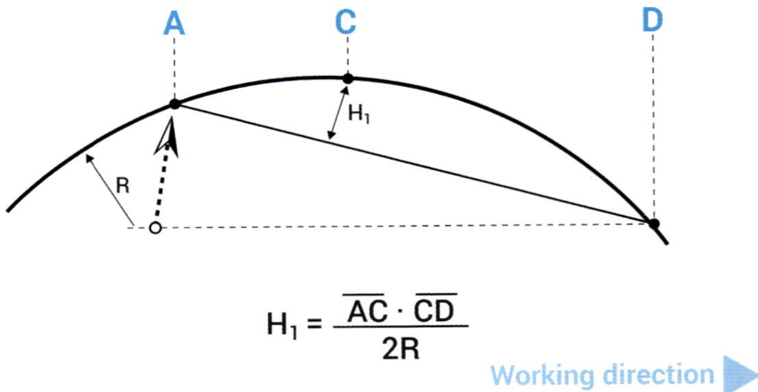

$$H_1 = \frac{\overline{AC} \cdot \overline{CD}}{2R}$$

Fig. 4-53: Diagram of 3-point compensation method

The principle of this method is as follows:

Point A is located behind the machine, on a track section that has already been lined. The front end of the chord, point D, is located by the alignment defect F_D. Point C is lined until the target versine at C meets the specified alignment. By specifying the target versine at C, C is put in a position that corresponds with the size of the desired curve radius. The residual defect F_R is derived from the ratio of the measurement point distances.

In the course of the lining work, point A will arrive at the residual defect and thus will influence the next measurement so that the deviation from the geometric target position will increase.

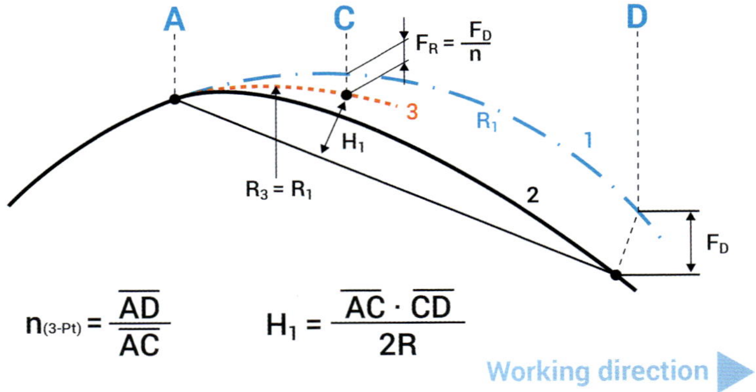

$$n_{(3\text{-Pt})} = \frac{\overline{AD}}{\overline{AC}} \qquad H_1 = \frac{\overline{AC} \cdot \overline{CD}}{2R}$$

Fig. 4-54: Reducing a defect with the use of 3-point compensation

4.5.3   Electronic compensation method

If the target geometry is unknown, the track geometry can be improved using the guiding computer of the tamping machine in combination with the chord measuring system.

Before track maintenance work commences, a separate measuring run (versine measurement in plan and elevation view) is carried out. Specific basic parameters can be the shape and length of the transition curves, radii and constant areas, linear areas and constraints. Based on these parameters, the guiding computer performs a compensation of the measured versines.

After the computation, the curve main data and slues are displayed on the screen. If the user agrees with the data, the determined geometry is stored. The track geometry is then corrected via automatic control using this data.

If track geometry documents are available, these alignment elements can be used as a basis for calculating the compensation procedure in the guidance computer.

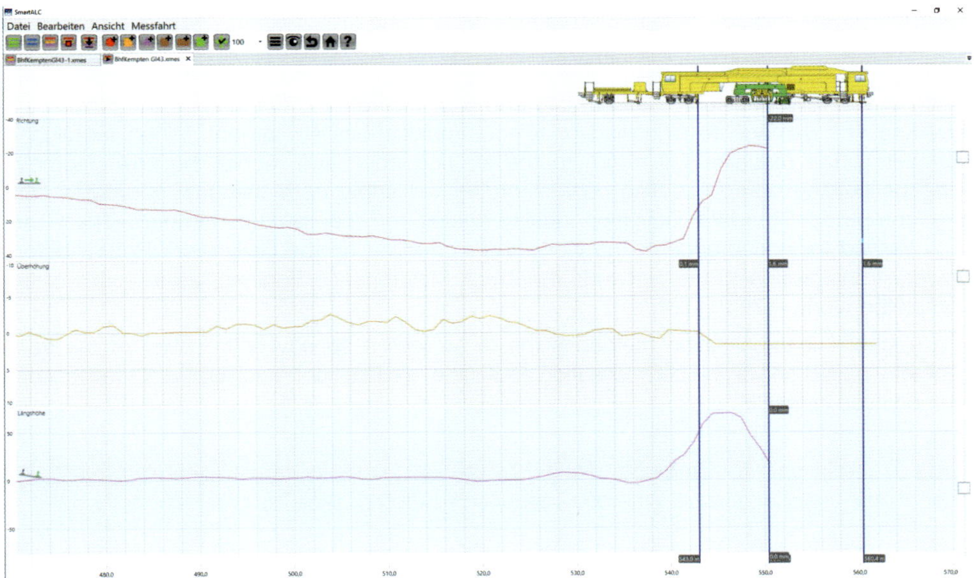

Fig. 4-55: Measuring run displayed at the guiding computer (example: Smart-ALC)

Track maintenance tamping in the UK uses two methods – geometry tamping and compensation tamping. Geometry tamping is similar to precision tamping (see 4.4 above) whereby a prior geodetic survey of the track has been carried out from which a geometric design with lifts and slues has been created that is input as a data file to the guiding computer of the tamping machine before work commences. Usually this respects the existing location of transition curves and cant values, and follows the maintenance limits for lifts and slues (see below). The track is not tied to fixed points at regular longitudinal intervals, however, there will be features that can act as a check as work progresses so that the new track position remains within the prescribed limits.

The predominant method of maintenance tamping of track in the UK is the electronic compensation method. The length of track to be tamped will have been prepared with the marking of sleepers to show key features of the geometry such as the top and bottom of transition curves; changes in cant every 5 mm and radii of curves. Obstructions and points of limited clearance where lifts or slues may not be possible will also be clearly marked. A starting point will be identified, preferably on straight track or in a constant curve but not in a transition, and the machine will commence moving towards the end of the worksite recording the horizontal and vertical versines on the machine's guiding computer. The operator will record key features such as transitions on the computer as the machine progresses along the track. On return to the starting point the machine supervisor will review the recording and set the computer to produce a smoothing of the measured horizontal and vertical versines by use of an algorithm, respecting any pre-determined limits of lift and slue that had been advised by the Network Rail Track Geometry Supervisor. For maintenance tamping using measurement and compensation, lifts and slues can (but generally do not) exceed 25mm. However limitations are mandated at +/- 25 mm under particular areas of normal overhead electrification (OHLE). Restrictions become more stringent through Close Tolerance OHLE and for listed tight clearances such as overbridges, tunnels, platforms and for passing clearances. The Track Geometry Supervisor will agree that the computed new alignment is compliant within the constraints they specified, before tamping can commence. During progress regular checks are made of the following: cross levels; line and top; clearances; damage to the permanent way and any damage to signalling or telecommunications equipment. When tamping is completed, the machine will return to the starting point and record the new geometry with the guiding computer. The new recording will show both the new geometry, cant and computed Standard Deviation quality values for the mileage tamped. These are both viewed for acceptance of the work and signed off on the Daily Work Return before onward transmission to the Network Rail Track Maintenance Engineer as a record of the completed work. [17] [21] [22]

The electronic compensation method helps to achieve a speedy optimisation of the track geometry based on versine compensation and is the predominant machine track maintenance method used in the UK.

Due to the inverse transmission function of the versine measurements and the resulting deviations, the compensation method cannot completely replace the precision method.

4.6   Correcting isolated defects (Spot Tamping)

If, during an inspection, defects are detected in the track geometry that exceed certain limits (e.g. SRA (DB) or SES (ÖBB, SBB)), these have to be corrected quickly in order to ensure safe railway operations and to avoid operating restrictions. In many instances, one of the first actions is to correct the track geometry by tamping, even if the cause of the track geometry defect may be found in other parts of the railway track, e.g. the ballast bed or the substructure.

Fig. 4-56: Caused by track geometry defects

A typical isolated defect would be approx. 3 – 20 m long, with deviations in the longitudinal level, twist and/or alignment or cross level. Spot tamping is an effective method to correct these isolated defects. A particular feature of the correction of isolated defects is that the track geometry (e.g. the longitudinal level) is not corrected to the target track geometry but only adjusted to the track geometry of the adjoining sections on either side.

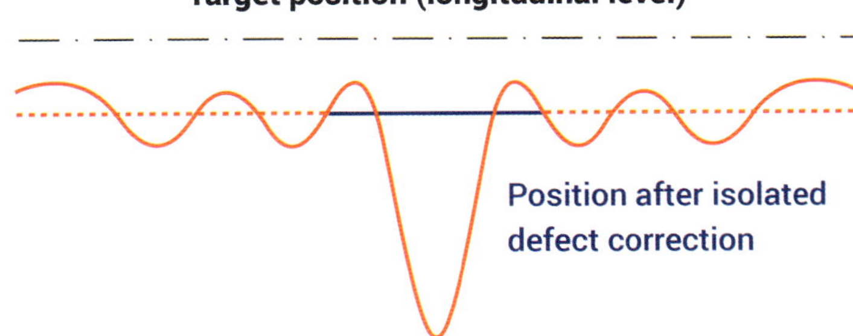

Fig. 4-57: An isolated defect: restored just below the target geometry

A tamping machine equipped for spot tamping carries out a measuring run over the area of the defect. This measuring run should start and end at least one machine length on either side of the isolated defect. The start point for measuring is marked on the track (e.g. a chalk line); this will make it easier to locate the starting position for the work activity and the final measuring run.

155

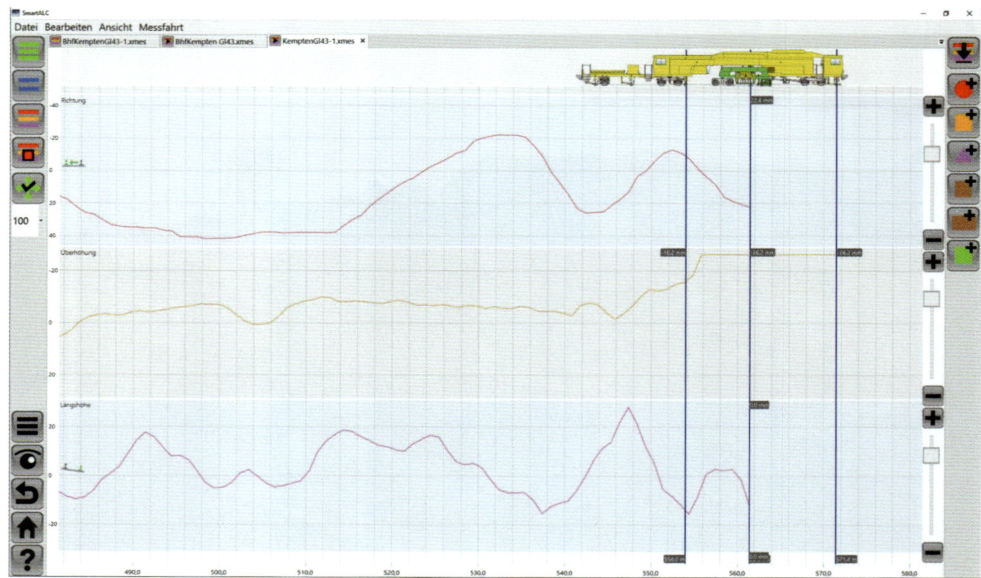

Fig. 4-58: Smart-ALC measuring run operator interface

During the spot tamping measurement, the versine of the longitudinal level of both rails, as well as the alignment and cant, are recorded. The computer system will then suggest one or more selection areas that the operator can change or confirm. For calculating the alignment and level compensation, the start and end points can be input separately.

If the suggested correction values meet the requirements, tamping can begin. Otherwise, there is the option to change the basic parameters of the compensation calculation (e.g. the start and end point, etc.).

When tamping, the start position has to be checked against the marking made and adjusted if required. A final measuring run needs to be carried out after or during the tamping run to check the work carried out. When correcting isolated defects, it is usually also necessary to sweep the sleepers and to re-establish the ballast profile.

4.7 Overlifting the track position as a maintenance process

The method of overlifting the track position is called design lifting or differential overlifting. Maintenance tamping aims to re-establish a good track geometry by removing level and alignment defects, to re-profile and compact the ballast bed. The aim is usually to achieve an actual track geometry that is as close as possible to the target geometry.

The use of design lifting creates an additional wear reserve in areas with strong or differential settlements. This will increase the interval to the next tamping work required in these sections and thus increase the sustainability and economic efficiency of tamping in these problem areas. Design lifting does, however, not remedy the causes of high settlement rates, but it improves the state of repair until the actual causes can be corrected.

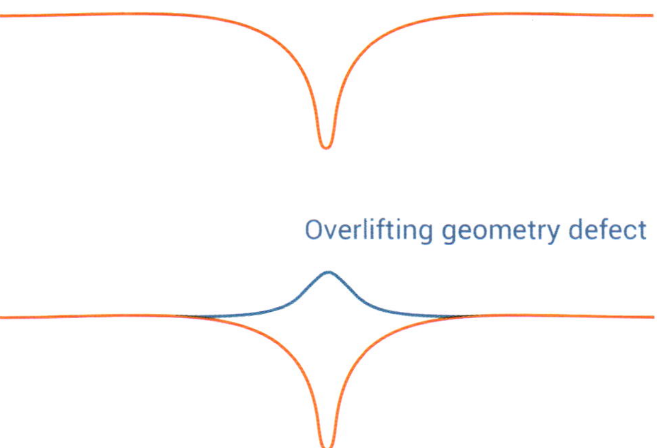

Fig. 4-59: Principle of design lifting

The overlifting values are calculated from envelope curves that are superimposed on the "normal" lifting values and can be of varying lengths. Broadly, this corresponds to subtracting the basic lift from the original lifting value in order to determine the actual size of the short-wave track geometry defect. This value is overlifted by D_F %.

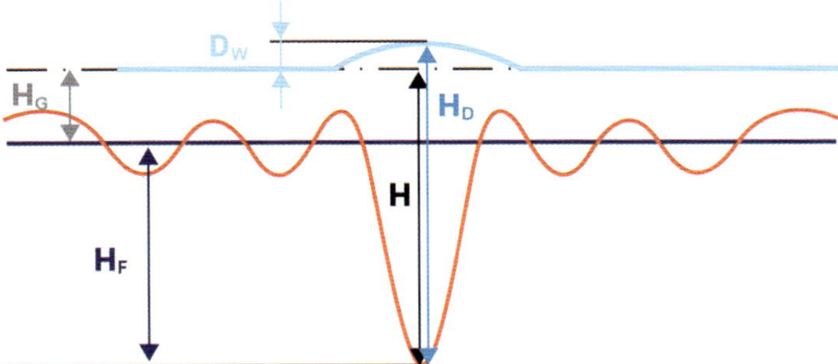

Fig. 4-60: Illustration of approximation formula for overlifting

Explanation of approximation formula:

$$H_D = H_G + H_F + \left(H_F \cdot \left(\frac{D_F}{100}\right)\right) \qquad (4\text{-}9)$$

H:	lifting value as a result of the classical target-actual comparison
H_G:	basic lift
H_F:	height (or depth) of the track geometry defect
D_W:	design lifting value
D_F in [%]:	design factor
H_D:	new lifting value for design lift

However, the following constraints must be taken into consideration:

– The design lifting value DW is limited to max. 10 mm.
– The design factor DF should range from 50 % to 70 %.
– The base length of the envelope is between 30 and 50 m.

This method is not to be used in turnouts and cross-overs, when tamping new track or renewed track, in platform areas or other areas where there are level restrictions.

The procedure of overlifting the track geometry is particularly suited for rectifying isolated defects or for tracks with strong longitudinal level defects.

4.8   Summary

In this chapter we have looked at the classic methods of measuring and surveying track prior to undertaking mechanised maintenance. Many of the manual methods such as the optic and the surveyor's level have been supplemented, either by the Total Station geodetic instrument or by the on-board measuring systems of modern on-track maintenance machines. However, the theory behind these new technology is still the same. This is why, despite all modern technology it is still crucial to understand the fundamentals of track surveying. In the next chapter we will see how these machines are used to correct track geometry defects.

5 The correction of track geometry

Fabian Hansmann and Richard Spoors

5.1 Key issues

This chapter deals with the whole tamping process as a work procedure with a focus on the underlying technology. An overview of the different tamping machines with their tamping units as well as the different work steps during their deployment will provide a step-by-step guide to sustainable maintenance of the track geometry.

- How do tamping machines differ from each other?
- Which work units are tamping machines equipped with and what tasks do they perform?
- What are the tasks of the tamping machine operators?
- What does one need to bear in mind before and after deploying a tamping machine?

5.2 Basic principles of track geometry maintenance

The railway track is capable of carrying high loads due to the interaction of its individual components (see section 2.4). Nevertheless, the constant impact of climatic conditions and traffic will result in wear of the system and leave its mark. In the case of track geometry, the acting forces will lead to more and more plastic deformations over time. The track geometry will increasingly deteriorate, resulting in track geometry defects. This degradation of the actual track geometry from the target track geometry is primarily due to deformations of the ballast bed as well as the subsoil and substructure. Fig. 5-1 shows a typical progression of the track geometry over the useful life of a mixed traffic line in Europe. [1, p. 65] The red line depicts the deterioration of the measured track geometry quality, the green line shows the maintenance activity carried out regularly to improve the track geometry, which will then deteriorate again at different rates.

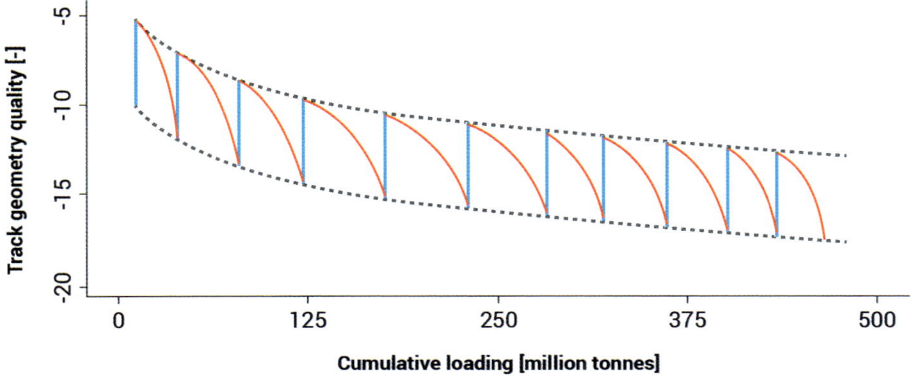

Fig. 5-1: Typical progression of track geometry development for a mixed traffic line in Europe, based on [1, p. 65]

Like with any other industrial assets, it will be necessary to maintain the asset, in this case the track and the track geometry, due to the wear arising from loading. According to EN 13306, maintenance is defined as the "combination of all technical, administrative and managerial actions during the life cycle of an item intended to retain it in, or restore it to, a state in which it can perform the required function". [2, p. 8]

Maintenance is an umbrella term for service, inspection, repair and improvement of an asset. [3] The principal objectives of maintenance can be defined as follows: [4, p. 13]

- prevention of system failures
- increase and optimum use of the service life of assets
- adherence to and improvement of operational safety
- increase of asset availability and optimisation of operating schedules
- reduction of faults
- forward planning of costs

As part of track maintenance, the track geometry is continuously measured and monitored using different methods (see section 4.2.3). This provides the basis for planning the maintenance and repair of the asset.

EN 13306 distinguishes between varieties of different types of maintenance. [2] In essence, track maintenance is predominantly carried out through preventive and corrective maintenance.

Preventive maintenance means taking actions to reduce the wear and failure probability of the system and its components. A general distinction is made between predetermined, condition based and predictive maintenance.

Corrective maintenance is an activity that puts an object back into a state in which it can perform its required function. Depending on the specified limits, the activity must be carried out either immediately or within a given time period.

The safe operation of the railway asset is always paramount and is the principal objective of every infrastructure manager.

In addition, aspects of sustainability, cost efficiency and ride comfort are also taken into consideration. Therefore, there is an increasing trend to move from purely corrective maintenance (reactive) to preventive maintenance (proactive). This is clearly shown when comparing the railway-internal intervention limits with those of the existing standard (see Fig. 5-2).

The differences reflect the various reasons for maintenance and furthermore, take into consideration the necessary contingencies. Track geometry defects not only have a considerable impact on the safe operation of railway assets, but also escalate the effects on the whole system (cp. with example of pothole in the road in section 2.4.1) and reduce the ride comfort. Therefore, it makes sense and is common practice for the railways of the DACH countries and the UK not to exploit the system reserves fully and to correct track geometry defects at an earlier stage.

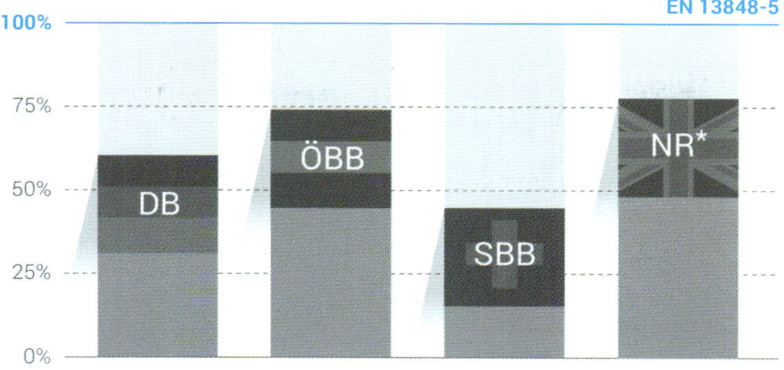

Fig. 5-2: Intervention limits for vertical track geometry D1 (zero to peak) in the railway-internal rules and standards as a per cent of the limits defined in standard EN 13848-5 [5] [6]

5.3 Correction of track geometry with a tamping machine

In any circumstance, a changing track geometry will eventually require a maintenance intervention. Europe-wide analyses show that the costs of maintaining track geometry dominate the overall cost of track maintenance. [7, p. 36] Therefore, both the quality and execution of work are of particular importance (see section 6.2).

In the DACH countries, a changing track geometry is corrected through tamping. Only at the end of a track's useful life is it difficult to establish a stable track geometry by tamping, due to the advanced wear of the ballast. The track geometry of such sections of line will deteriorate again quickly after tamping, and the intervals between the interventions become shorter and shorter. In this case, cleaning the ballast will lead to a lasting improvement of the track geometry. [8, p. 69] According to EN 13231-1 [9], ballast bed cleaning is technically not considered as maintenance, due to the partial replacement of ballast. Therefore, it will not be considered further here. In the UK tamping machines and the Stoneblower are used to maintain track geometry.

In the 1980s, alongside the classic tamping machine, British Rail developed a maintenance machine it called the Stoneblower. [10] In simple terms, the machine is an automated mechanical exploitation of the classic measured shovel packing of track using graded chippings pneumatically injected under the sleeper in measured quantities through inserted tubes. Over thirty years, experience with the machine has shown that the life of older track with poor ballast conditions can be extended. Stoneblowing, however, should not be regarded as an alternative to tamping. In the UK, the two methods are seen as complementary. Tamping has been proven to be very successful in correcting and sustaining track geometry and is used on railways worldwide. However, once the ballast bed starts to become heavily fouled with fines, the resulting track geometry after tamping is no longer sustainable. A reduced interval of tamping interventions becomes necessary to keep the track geometry at a safe level. Refurbishment by ballast cleaning is now required to ensure the track's sustainable operation and improve the asset condition. In the UK, utilisation of the Stoneblower, particularly on secondary lines, enables the track to be safely maintained for a little longer before refurbishment is required. Once a section of track has been selected for treatment by stoneblowing, it is no longer suitable for tamping as the properties of the ballast bed (e.g. permeability) have been changed by the introduction of fines. This chapter will focus on tamping machines.

5.3.1　Different types of tamping machines

Numerous individual demands are placed on tamping machines. This has resulted in the development of different types of tamping machines by European manufacturers over the years. Their names vary between the manufacturers and the guidelines of the railways.

The following classification is an attempt to divide tamping machines into main individual groups, despite those differences. Some rules and standards may, however, deviate from this classification. Generally, tamping machines can be divided into the following five groups:

- **Plain line tamping machines**
  - single-sleeper tamping machines
  - two-sleeper tamping machines
  - three-sleeper tamping machines
  - four-sleeper tamping machines
- **Universal and switch tamping machines**
  - single-sleeper tamping machines
  - two-sleeper tamping machines

  Depending on the machine type, modern tamping machines can work up to four sleepers simultaneously. These machines can achieve peak outputs of over 2,600 m per hour.

  In addition to the number of sleepers to be worked on, plain line and universal tamping machines can be designed as cyclic or continuous action machines.

  With cyclic action tamping machines (see Fig. 5-3) the whole machine must be braked to perform the work, i.e. for lifting, lining and compacting the track. After that, the whole machine will move forward, brake and work on the next section.

Fig. 5-3: Cyclic action two-sleeper tamping machines: B 45 D (MATISA, left) and 08-32/4S (Plasser & Theurer, right)

With continuous action machines (see Fig. 5-4) the machine frame will continue moving while the tamping bank, which is mounted on a freely moving tamping satellite, can work independently. This makes it possible for the machine to move continuously during tamping if the conditions are suitable and so achieve constant higher outputs while protecting the machine.

Fig. 5-4: Continuous action multi-sleeper tamping machines: B 50 D (MATISA, left) and 09-3X (Plasser & Theurer, right)

Universal tamping machines are suitable for the maintenance of turnouts. The special geometry and component sensitivity of turnouts require high adaptability of the machine. Individual work units of universal tamping machines can be adapted more specifically to these special requirements.

– **Road-rail tamping machines or self-loading tamping machines**
These are very compact, fully functional tamping machines. As they are easy to load and very flexible, they can be readily available, particularly on sections of line that are not connected or where the machine cannot transit by rail for other reasons.

Fig. 5-5: Self-loading tamping machine B 38 C (MATISA, left) and universal tamping machine with road travelling gear UST 79 S (Plasser & Theurer, right)

– **Small tamping machines**
There is no clear definition of small tamping machines. RIL 824.3401 describes small tamping machines as "self-propelled or not self-propelled machines that can be used for tamping ballast under the sleeper and compacting it". [11, p. 1] Unlike large tamping machines, small tamping machines will often have no lifting and lining facility; this task will need to be carried out using suitable auxiliary equipment. Even if machines have suitable lifting equipment, the track lift to be achieved is limited, especially in the case of heavy concrete sleepers, due to the low counterweight of the machine.

Fig. 5-6: Small tamping machine Minima 1 (Plasser & Theurer, left) and ROJACK rail lifting device (ROBEL, right)

– **Attachment tamping banks and manual tamping units**

In addition to tamping machines, attachment tamping banks and manual tamping units are used in some instances. However, their use is work-intensive and quality-sensitive and is limited to:

- isolated defect correction that requires immediate temporary intervention
- tamping of partial sections that cannot be tamped with a machine due to different installations
- transition ramps on construction sites
- temporary tamping on small-scale local construction sites so that the track can be used by trains

Even though their use is limited, manual tamping units are still an indispensable piece of equipment for track maintenance.

Fig. 5-7: Manual tamping units (vertical tampers) ROTAMP in practical use (ROBEL)

5.3.2   Choice of machine type for the UK

Each Infrastructure Manager has a policy for the deployment of on-track machines and, where the service is bought in through contracts, it is the contract that determines the particular type of machine that is procured and used. Network Rail let a small number of seven year contracts for services to provide maintenance and renewal tamping contract shifts in geographic areas in 2018. This allowed the successful companies to invest in new machines from Matisa and Plasser and Theurer. Their choices included the B66 UC from Matisa and the Unimat 09-4x4/s Dynamic from Plasser and Theurer (see Fig. 5.8). These continuous action machines are flexible; they can be used on maintenance and renewal of either plain line or switches and crossings. This is considered to give an advantage in logistics planning as machines spend less time traveling from site to site, albeit at a reduction of output per hour when plain line maintenance tamping is carried out when compared to a dedicated plain line machine. This highlights the choice of machine type which will be influenced by many parameters, for example, life cycle costs, performance values, annual output of track tamped, travelling speed between sites, planning logistics and possession lengths. Each machine has the respective company's guiding computer; CATT from Matisa and Smart ALC from Plasser and Theurer. Both these systems are designed to support tamping with either known or unknown geometry and comply with Network Rail's specification for post tamping quality assurance described in Chapter 6.

Fig. 5-8: Continuous action tamping machines for the UK: Matisa B66 UC left and Plasser Unimat 09-4x4/4s Dynamic right. [12]

5.3.3 Work units of a tamping machine

In the course of a construction project, the tamping machine will have various supplementary work units at its disposal for establishing the required track geometry. These work units will differ from each other, depending on the type of work, as some of them are very specifically aligned to the requirements of the machine. As a general guideline, five main groups can be distinguished:

- lifting-lining unit
- tamping bank
- ballast shoulder consolidator
- chord measuring system for carrying out the work
- chord measuring system for documentation/control measurement

Even though different tamping technologies differ concerning individual tamping parameters, the process of correcting track geometry defects is de facto always the same. On the basis of the specified correction values, the tamping machine will lift (vertical position) and align (horizontal position) the track to the specified geometry to establish the required relationship between the two rails. To achieve this, the tamping machines use lining rollers and lifting clamps (see Fig. 5-9). They are engaged on the side and underneath the rail head to move the track into the desired position.

Fig. 5-9: Open (left) and closed (right) lifting-lining unit showing the roller clamps

Universal and switch tamping machines are usually also fitted with lifting hooks. These are useful especially if the lifting clamps cannot grab the track panel from underneath the rail head, e.g. if there are fishplates. Lifting hooks are more flexible and can, if necessary, grab the rail at the foot, or the check rail or wing rail in the area of the switch frog point. Some work units will also allow the operator to control the closing mechanism of the lifting clamps. Amongst other things, the operator will be able to close the lifting clamps shortly before lowering the tamping bank so that the closed clamps do not accidentally damage any fitted obstacles when the machine is moving forward.

Fig. 5-10: Lifting hook on a universal tamping machine

The machines make use of their integrated chord measuring systems based on the 3- or 4-point method to establish the desired longitudinal level and alignment. During machine setup, the measuring axles are lowered onto the track and pressed against the reference rail corresponding to the alignment of the track. In addition to the correct alignment and level of the track, the fitted measuring system can also determine the cant of the track. The chords used are laser or steel chords (see Fig. 5-11).

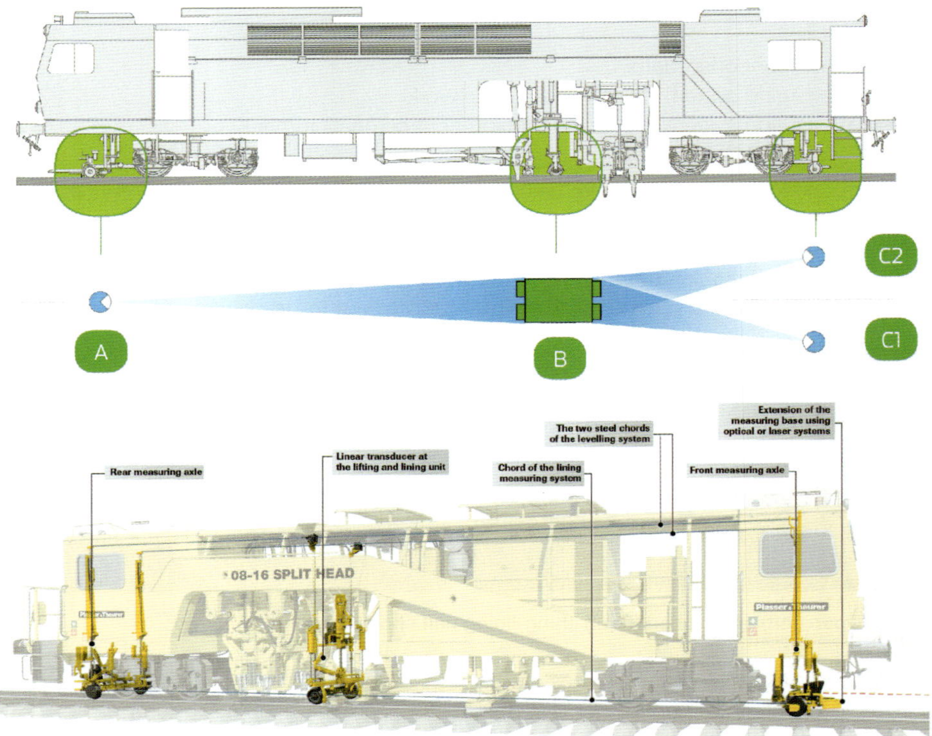

Fig. 5-11: Tamping machines use steel or laser chords for correcting the track geometry

Generally, the infrastructure manager defines as reference rail the outer rail for alignment and the inner rail for lifting. In Germany, Austria and the UK, the target geometry and correction values refer to the reference rail. In Switzerland, the track centreline serves as the basis for the longitudinal level. This must be taken into consideration for tamping machines that are deployed in Switzerland. The alignment values are, however, not affected by this.

As a result of the geometry correction, a void will occur at the sleeper end and especially under the sleeper. To permanently fix the track into its new position, these voids have to be filled and compacted with the use of the so-called squeezing process. This process is made up of three parts: 1. Penetration, 2. Squeezing, 3. Raising.

In parallel to the lifting and lining process, the tamping tines fitted to the tamping bank penetrate the ballast crib (penetration) and perform a closing movement towards the sleeper centre (squeezing). The ballast moves in the same direction; fills the void and forms a compacted bearing surface under the sleeper from the vibrations of the tamping tines. After squeezing has been completed, the tamping bank will be raised again (raising); this completes the squeezing process.

Fig. 5-12: Graphic representation of the different phases of a squeezing process: penetration, squeezing and raising

Several squeezing procedures (also called interventions) can be performed per sleeper before the machine completes the tamping process by moving to the next sleeper or group of sleepers (in the case of multi-sleeper tamping machines).

The tamping tine has a conical tine shaft that connects it to the tamping bank. On its underside, the shaft widens to an armoured tine plate that is responsible for moving and compacting the ballast. During the squeezing process, it is moved against the ballast and therefore is subject to particularly high requirements. Its exact shape varies from manufacturer to manufacturer and from machine to machine. Whether its specific shape meets the railways' rules and standards will be checked when awarding approvals for work and deployment. Railway-internal rules and standards may specify in special cases that the tamping tines have to be checked before each deployment (e.g. with a limit value gauge) and replaced, if necessary. [13]

Fig. 5-13: Typical shape of a tamping tine and its fixing to the tamping bank

Ballast sleeper end consolidators compact the ballast at the shoulder and are used adjacent to the tamping bank or slightly behind it. In this way, the void created by the lateral displacement of the track panel between the sleeper end and the shoulder ballast can be reduced during the tamping process. The track's lateral resistance, reduced by the track geometry correction process, will be restored. It may be necessary to take additional measures (see section 6.2) before the final re-opening of the track.

Ballast shoulder consolidators cannot be fitted to tamping machines in the UK due to the reduced structure gauge.

Fig. 5-14: Ballast shoulder consolidator for compacting the ballast at the sleeper ends

Two machine operators generally control the tamping operation of a plain line tamping machine. One operator (the front tower operator) sits in the front cabin of the tamping machine, sometimes called the front wagon, and from this position, he monitors all track geometry parameters before and after the tamping work (see section 5.3.4). In addition, the front tower operator has an unrestricted view of the track ahead from the front wagon and thus is able to warn the second operator of any tamping obstacles.

The tamping cabin is located right above the tamping banks and the lifting-lining unit. From here, the second operator checks the smooth running of the tamping process and responds to any special situations using the lifting-lining unit and, where fitted, the ballast shoulder consolidator. Compact tamping machines will often be controlled by only one operator, who will then have to take control of all the work units. Universal tamping machines may require an additional operator who is in charge of parts of the tamping bank or the lifting-lining unit to meet the strict requirements of turnouts.

5.3.4   Preparatory work for a sustainable tamping process

In order to ensure the sustainability of a tamping activity, a number of preparatory tasks have to be carried out before the work commences, in addition to the measuring work mentioned. Depending on the contractual arrangements, these activities will be the partial or complete responsibility of the contractor or the infrastructure manager.

If for cost reasons the output quality of the tamping work is reduced due to economies in the preparatory work, this will have a negative impact on the sustainability of the tamping work and in turn reduce the interval to the next intervention. Thus, the best possible output quality will not only reduce the overall cost but also retain a high quality standard of the assets. [14, pp. 32ff] Preparatory work includes the following activities:

–   Repairing the drainage (e.g. cleaning the trenches, the underground drainages, etc.)
    A well drained and well ventilated ballast bed will ensure the best possible result of a track geometry correction; on the other hand, waterlogging can have a negative impact on the result.

Fig. 5-15: Wet spots suggest advanced wear of the ballast or fines rising from the subsoil

–   Removal of wet spots or mud spots
    Wet spots occur as a result of increased ballast abrasion, mud spots on the other hand as a result of fines arising from the subsoil. Both phenomena in an advanced state can be clearly seen on the ballast surface (see Fig. 5-15) and, due to the high proportion of fines, make it almost impossible to carry out a lasting correction of the track geometry.
–   Checking that connections are secure between fastening systems, rails and sleepers and replacing any damaged sleepers
    Damage to the track panel can make it impossible to re-establish the correct track geometry. If the connection between the components is no longer sufficiently secure, the rail may detach from the sleeper during lifting and lining. The result will not only be of insufficient quality but will also pose a high risk to the safety of the track maintenance work.
–   Replacing defective rail pads for a sustained reduction of the impact on the whole system
–   Adjustment of the track gauge
–   Repair of rail and insulated joints as well as maintenance of the rail surface to minimise the impact of wheel/rail forces
–   Obtaining and checking the target data
–   Pre-measuring the track geometry
–   Determination of ballast requirement and depositing the ballast
–   Dismantling the cover of level crossings
–   Loosening and dismantling anti-creep protection
–   Loosening and dismantling sleeper end plates
–   Loosening fastenings over a length of approx. 10 m at the start and end of bridges without a continuous ballast bed
–   Dismantling train detection systems, e.g. Indusi, rail contacts, treadles etc.
–   Dismantling check rails and guard rails as well as other installations if they are within the working area profile
–   Removal or the clear marking of other tamping obstacles

5.3.5   Track geometry data input

Once a tamping machine has reached the tamping area and has been correctly set up, it will require the necessary track geometry correction data and the cant target value specifications to be able to start work. This correction data contains the values for lifting and lateral displacement of the track. It will either be transmitted separately to the machine in digital format or determined directly using the compensation method where a measurement run over the track to be tamped is performed. For carrying out the compensation method with known track geometry, the front tower operator has to input the target values (track geometry parameters) for alignment, cant and level of the track position before starting work. These target values are to be taken from the track measurement plan (also called track layout plan) or are supplied by the client in a different format.

Fig. 5-16: View of the track from the front cabin

Modern tamping machines make use of different types of guiding computers to determine the correction values themselves, if necessary, to process them and transmit them to the work units. Here, they make use of the 3- or 4-point method (levelling and lining measuring system), taking into consideration the machine-specific conditions, such as chord division, correction values (not to be confused with the correction data) or the lowering dimension. The procedure adopted will depend on whether the tamping process uses the precision or compensation method.

If tamping machines use the compensation method to correct the track geometry, they will have to consider the appropriate **correction values** for curvature changes due to the chord measuring system. The guiding computer can do this automatically, and also most tamping machine manufacturers provide correction value tables for common geometry.

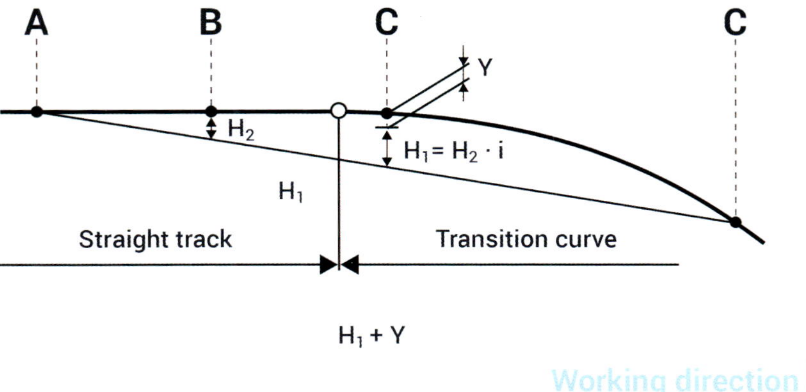

Fig. 5-17: Correction value 'V' in the transition curve using the 4-point chord measurement method

Since in a transition curve the chord moves towards the centre of the curve as the radius reduces, the slueing values for the machine will be distorted. Therefore, in a transition curve, an **error dimension**, which serves the purpose of a correction value, is adjusted gradually. This only applies to canted curves or transition curves.

The machine control system can be upgraded with various modules. These manufacturer-dependent programs will make it possible to perform spot tamping, the direct input of geometry data from recording cars, design tamping or other special applications. Apart from the automatic input via the guiding computer, the machine can also be guided via laser beams or optical equipment such as Plasser and Theurer's curve laser.

The front tower operator can intervene in the control program of the tamping machine at any time to make adjustments to the specifications. It is also possible to input correction values manually into the system. This will be required if the correction values are not transmitted to the machine digitally but are marked on the track (see Fig. 5-18).

Fig. 5-18: Lining and lifting values written on a sleeper

In general, when inputting lifting values, attention needs to be paid so that applicable limit values are not exceeded. Exceeding permissible lifting values has to be avoided for two reasons:

1. Rail stresses

   Stresses in the rail occur in the course of track geometry correction as a result of bending. Depending on the rail profile and the distance between the bogies of the tamping machine, these tensions will be of varying values. To avoid critical manifestations of rail stresses that may result in track buckling, the lifting values have to be limited.

2. Consolidation quality of ballast

   Large depths of new ballast cannot be correctly consolidated. Therefore, in the course of new construction or track renewal, the DACH countries have moved to consolidating the ballast in layers (see Fig. 5-19). However, it is for this reason also that the lifting values have limits for general maintenance work.

   In the UK, when category 1 and 1A track is renewed, and the ballast bed is excavated, Network Rail permits the ballast bed to be installed in one layer, with both bottom and top profiles controlled by beacon laser. Consolidation is with approved vibrating plates and should be such as to safely open the line, sustain traffic at the required speed and achieve the track geometry standard. On lower category track, with line speed below 170 kph, the track may be renewed by being lifted through the ballast in layers. The maximum lift is 100 mm. [15]

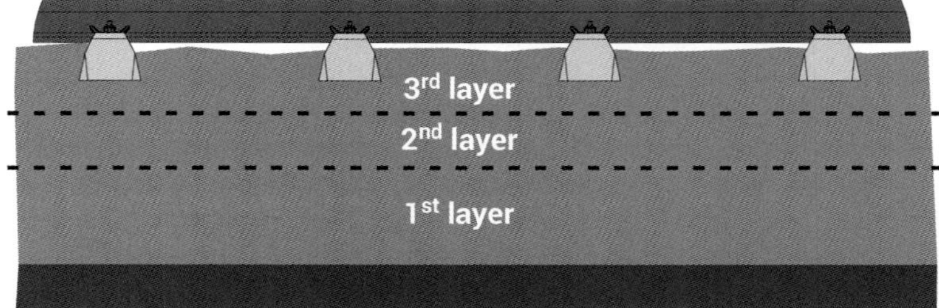

Fig. 5-19: Building up the ballast bed in layers to obtain the required consolidation

Fig. 5-20 provides an overview of the currently applicable maximum values for lifting a plain line track panel, resulting from a combination of rules and standards [16] – [19] and practical experience. To evaluate if a tamping process has been carried out correctly or not, the acceptance parameters listed in section 6.2.3 will be used. The number of interventions defines how often the tamping bank penetrates the ballast in one ballast crib and performs a squeezing movement.

DB	ÖBB	SBB	NR
New layer	**New layer**	**New layer**	**New layer**
Layer 1 from formation	Layer 1 from formation	Layer 1 from formation	≤ 100 mm
≤ 60 mm: 2 interventions	≤ 150 mm: 2 interventions	≤ 60 mm: 2 interventions	
		≤ 100 mm: 3 interventions	
Layer 2	Layer 2	Layer 2	
≤ 25 mm: 1 intervention	≤ 80 mm: 2 interventions	≤ 60 mm: 3 interventions	
≤ 30 mm: 2 interventions			
Layer 3	Layer 3	Layer 3	
≤ 25 mm: 1 intervention	≤ 30 mm: number of interventions as required	≤ 60 mm: 2 interventions	
Maintenance	**Maintenance**	**Maintenance**	**Maintenance**
≤ 25 mm: 1 intervention	≤ 60 mm: number of interventions as required	≤ 60 mm: 2 interventions	≤ 25 mm: 1 intervention
≤ 40 mm: 2 interventions			

Fig. 5-20: Overview of the maximum lifting values common in the DACH countries [16] – [19] and in the UK [20]

Special requirements apply to steel sleepers in Germany and Switzerland which mostly specify an additional intervention.

In addition to maximum lifting values, minimum lifts are also specified. In the DACH countries, these are usually between 15 and 20 mm. From experience, this minimum lift is necessary to create sufficient space for the ballast stones to move and re-consolidate as a result of the squeeze action. This value is significantly influenced by the distribution of the stone size, the minimum of which is defined at 22.4 mm (see section 2.6.5).

Run-out ramps have to be created, in order to have as smooth a transition as possible between tamped and untamped areas and without exceeding safety limits. Similar to cants, run-out ramps serve the purpose of grading the alignment as well as the rail levels between tamped and untamped areas.

Here, the basic track design rules have to be adhered to and/or limit values obtained from rules and standards. For example, RIL 824 2200 differentiates the ramp gradient depending on the speed. [21] Accordingly, the length of the ramp is 2000 times (v > 160 km/h), 1000 times (80 km/h < v ≤ 160 km/h) or 600 times (v ≤ 80 km/h) the last lifting value.

The ÖBB rules and standards pursue a similar approach. They define the length of the ramp depending on the speed and the last lifting value as follows: [22]

- $v \leq 120$ km/h   1 : 800
- 120 km/h $< v \leq$ 160 km/h   1 : 1000
- 160 km/h $< v \leq$ 200 km/h   1 : 1500
- 200 km/h $< v$ 1 : 2000

In Switzerland, the transition is called a connection ramp and is executed at 0.5 per thousand (1:2000) in curves with radii bigger than 125 m and on straight line sections and at 1 per thousand (1:1000) in curves with radii of 125 m or smaller. [19]

There are no specific requirements for run-out ramps in the UK.

5.3.6   Controlling the tamping process and its parameters

The actual tamping process is controlled by the operator from the tamping cab adjacent to the tamping bank. Thanks to his clear view of the tamping and lifting-lining unit, the operator is in a position to guide and monitor these units correctly. Furthermore, the machine control system will provide an overview of the key geometry data so he or she can respond to any anomalies at any time.

Fig. 5-21: View of the tamping bank and the lifting-lining unit from the tamping cab

Some installations, e.g. axle counters, may get inadvertently damaged by the lifting-lining unit. To reduce this risk, such obstacles need to be clearly marked before tamping so that the operator in the tamping cab can see them in good time. This allows the operator to adjust the work units to the requirements in the best possible way. For example, the lifting clamps in critical areas will only close when the tamping bank is lowered so that no installations are damaged by the closed clamps when the machine is moving forward. Ideally, if at all possible, all obstacles should be removed before work starts. Generally, the work units are designed such that obstacles such as fishplates will push open the roller lifting clamps without damaging them, and so neither the track nor the machine will suffer any damage. The variably adjustable height of the roller lifting clamps can be adjusted to the requirements of different rail profiles. In the case of special installations, where it is impossible for the roller clamps to be activated on one side or in specific areas, they can be controlled individually.

The tamping cab operator has the task of positioning the tamping unit directly above the centre of the ballast crib. In the case of variable sleeper spacing or double sleepers, the opening width of the tamping unit has to be adjusted so that the track components do not suffer any damage.

Upon confirmation by the operator, the tamping unit will automatically lower and start with the specified squeezing procedure. Here it is critical that the tamping tines do not damage the sleeper or other components during penetration and the subsequent squeezing process. With plain line tamping machines, 16 tamping tines per sleeper, divided into two groups of eight on either side of each rail, will penetrate the ballast (see Fig. 5-22).

In the UK the railways to the south of the river Thames are nearly all electrified with a third rail system. Therefore, tamping machines have to have special tamping banks whereby the two outer tines are removed to avoid striking the third rail. As there are a variety of machines in use in the UK from both Plasser and Theurer and Matisa, Network Rail provides a guide to each machine's characteristics such that the site supervisor will be familiar with the position of the tamping bank operator and his field of vision, in order to complement the marking up of track and to minimise the risk of damage to track features during the tamping process. [23]

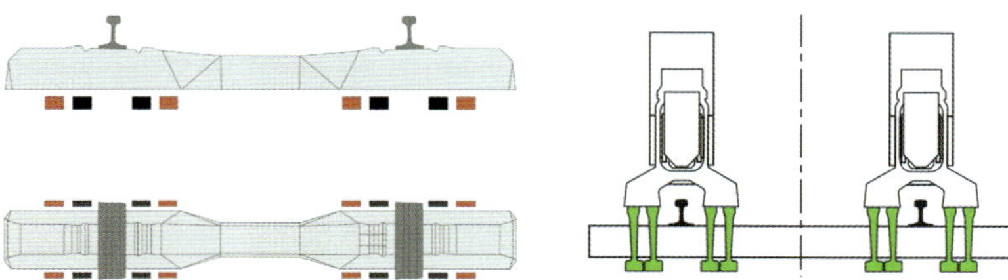

Fig. 5-22: Schematic diagram of tamping tines on a single-sleeper tamping machine penetrating the ballast

In order to prevent damage to the bottom edge of the sleeper and to perform the best possible compaction of the ballast under the sleeper, the operator has to know the construction height of the track and input it into the machine control system. The construction height of the track is derived from the sum of the heights of the parts - rail, rail pad, sleeper and, if present, under-sleeper pads. Furthermore, according to EN 13231-1, the penetration depth of the tamping tines has to be selected such that the top edge of the tamping tine comes to lie 15 to 30 mm below the sleeper bottom edge (including the under-sleeper pads). [9] When taking into consideration the limit values in the DACH countries and the UK, this will achieve the required output values. If the track conditions change in the course of a section of track, these changes need to be marked clearly for the operator. This way the operator will be able to adjust the values on the machine in good time. The DACH countries require the following distances between sleeper bottom edge and top edge of tamping tine:

- DB: 20 mm [21]
- ÖBB: 15 – 20 mm [24]
- SBB: 20 – 30 mm [25]

In the UK, Network Rail lists tamper tine depths for all plain line track types and S&C bearers in Standard NR/L3/TRK/3241 – Marking of Track for Tamping Machines. For the standard G44 sleeper and Fastclip fastenings, the depth is 380 mm which gives an equivalent distance within the range of 15 – 20 mm. As it is possible during a tamping shift for two different track types (sleeper/fastening/rail) to be encountered, the change point is marked on the sleeper, e.g. "TD 365".

Once the tamping tine has penetrated the ballast, the squeezing process, which is carried out by each pair of tamping tines separately, will start. This is the so-called non-synchronous constant pressure tamping principle. The tamping tines will keep squeezing until the set counter-pressure is achieved (between 95 and 115 bar, depending on the setting and the tamping machine) or the specified time limit has elapsed. This has the advantage that homogeneous compaction of ballast can be achieved under both rails.

The tamping tines compact the ballast under the sleeper through vibration and the squeeze direction; the frequency is between 35 and 45 Hz, depending on the technology used. In principle, elliptical and linear vibration movements are used in Europe. They differ from manufacturer to manufacturer with regard to the operating principle and the effect. While linear tamping technology makes use of frequencies at 35 Hz, the elliptical principle relies on a higher frequency of 42 Hz. It is important that sufficient energy for compaction is transferred to the ballast through a sufficient amplitude (4 – 5 mm) during squeezing. This amplitude will also have to be transferred to the ballast in difficult conditions, since the ballast bearing surface will otherwise not be compacted to a sustainable quality.

The tamping tine performs a vibration movement during squeezing. It moves in one motion forward and then backward in equal distances from its central start position. Once the tamping tine reaches its central start position again, it will have completed a full vibration. The number of vibrations that a tamping tine performs per second is its frequency, and the deflection from its central start position is its amplitude.

Due to the vibration of the tamping tine, it will only come in contact with the ballast under the sleeper for a fraction of the squeezing movement. Therefore, it is all the more crucial for a successful result of the compaction to select sufficient time for squeezing (squeezing time). EN 13231-1 recommends 1.0 to 1.2 seconds. [9] This value will depend on the implemented lifting value and the compaction characteristics of the track ballast. Specific values will be supplied by the client but should not be below 0.8 seconds. This recommendation applies to Germany [21] and Austria. In Switzerland, the squeezing time is called "closing time" and is specified at 1.0 seconds [25]. In the UK a specific value is not defined, and the setting is the responsibility of the machine supervisor, taking into account the range of lifts and slues; i.e., the larger the lifts bigger the setting. This is to allow the machine to complete the cycle prior to the squeeze action. The squeeze pressure should be adjusted by the operator dependent on the track formation and the maximum pressure applied without lifting the track by the squeeze action.

In general, the behaviour of the track should be observed during squeezing. If the ballast starts flowing out beyond the sleeper end or the sleeper is moving, the squeezing time and other tamping process parameters (e.g. the squeeze pressure or the number of interventions) will have to be adjusted.

If the ballast is not compacted correctly, the sleepers will not lie wholly on the ballast bed after tamping, and voiding layers will form as a result. This will result in increased impact stresses when trains are travelling over the area, which corresponds to accelerated wear of the ballast. After some time, the consequences will be seen on the surface of the ballast (see Fig. 5-23).

On steel sleepered track, once the tamping process is completed, the tamping tines are reset, and the unit is raised. If the tamping bank is raised too early, the edge of the steel sleeper may bend up and cause damage to the sleeper.

Fig. 5-23: Voiding in the ballast layers will result in higher ballast wear

If the tamping machine has ballast shoulder consolidators, they can be used automatically as the tamping bank is lowered. This is done automatically, and the operator only has to ensure that the use of the consolidators does not cause any damage to the railway infrastructure (e.g. in the area of level crossings) or the machine. This completes the actual tamping process, consisting of lifting, lining and compacting.

5.3.7   Spot tamping

When a short wavelength track geometry irregularity occurs in either rail or both rails the default option to effect a repair and prevent a more significant problem would be to organise a manual repair with hand tools. A more sustainable repair using a tamping machine has been developed and is particularly economic where several short locations can be treated in a single shift. The solution is new software called "Sprinter" (Guided Spot Tamping) [26] for the guiding on-board computer that produces a sequence of track lifts that are applied independently to the rails with the defects. A comparison to this system is the use of independent sighting boards on each rail. Using this innovation, tamping machines can rectify single faults up to a chord of 20 metres in length, even in transitions. (see a more detailed description in section 4.6 of Chapter 4)

5.3.8   Parallel tamping

In the UK, Network Rail has developed parallel tamping whereby two machines work as one in a master/slave configuration to tamp switch and crossing units. In chapter 2, Fig. 2-66 showed how the narrower distance between tracks in the UK (1970 mm) has created a design of crossover with long bearers between respective crossings. By placing one machine on each track and connecting their tamping banks and guidance system with an umbilical cord the machines are able, with an uploaded tamping file, to correctly repair geometry defects in the crossover as if they are a single unit.

Fig. 5.24 Parallel tamping switches and crossings at Hanslope, UK, on a wet Bank Holiday weekend

5.4 Dynamic track stabilisers and consolidators

In the DACH countries, tamping machines are often accompanied by dynamic track stabilisers (DTS) or consolidators to increase the sustainability and safety of the tamping process even further. Modern machines can combine the whole tamping and stabilisation/consolidation process in one machine.

In the UK independent, DTS machines have been phased out and replaced by integral units within the tamping machine.

Tamping acts on an existing ballast bed that has been heavily compacted by traffic to return the track panel to the desired geometry. The controlled squeeze action compacts the bearing surface under the sleeper as best as possible. Nevertheless, limited voiding will remain in the ballast crib after lifting the tamping tines, i.e. the ballast bed surrounding the sleeper will be loosened as a result of track geometry correction (see Fig. 5-25). This will have a significant effect on the tamping result:

- The unfilled voids will reduce the lateral displacement resistance of the track panel to a level that will restrict operations.
- The tamped track is subject to particularly high initial settlements in the first few months. Like with every compaction process, after initial loads have been applied, there will be shifts in the ballast structure until the ballast bed has sufficiently consolidated. Due to these applied loads, the initial track settlement will be irregular, which will increase dynamic excitation. In turn, this will reduce the interval to the next tamping process.

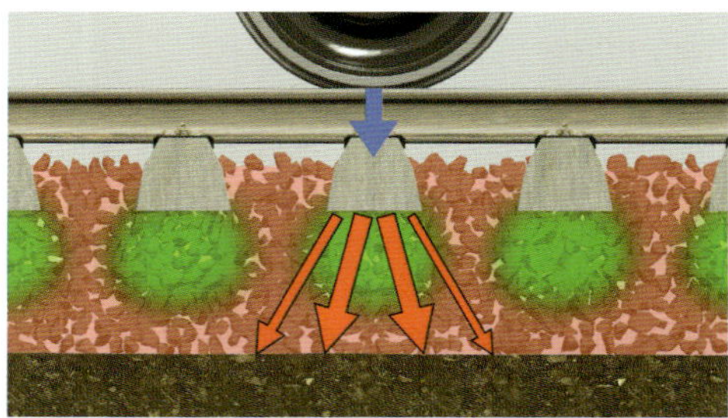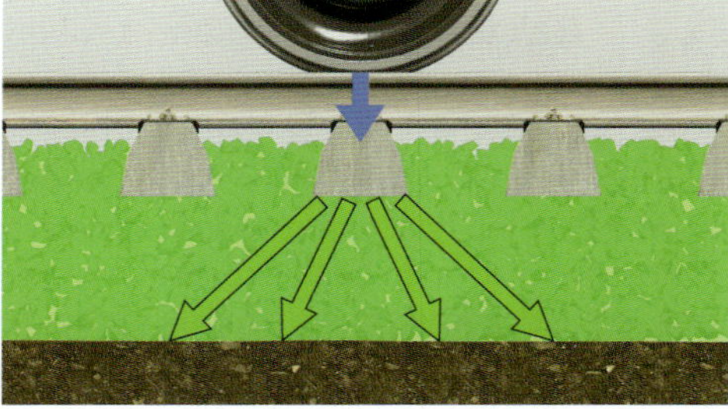

Fig. 5-25: Schematic diagram of the effect of dynamic track stabilisers

DTSs and consolidators have the aim of keeping the influences of voiding and high initial settlements on tamping quality as low as possible and to pre-empt them in a controlled manner. While the dynamic track stabiliser achieves this effect through horizontal excitations applied to the track panel, consolidators consolidate the ballast crib and try to achieve the same effect in this way. This makes it possible to influence the consolidation process in a targeted manner and to increase the lateral displacement resistance to a level where most reduced line speeds will no longer be necessary after the disturbance of ballasted track.

If neither a DTS nor a consolidator is deployed after tamping, it may be necessary to travel on this section of track at reduced speeds, depending on the type of track and the work undertaken. This reduced speed running will consolidate the ballast and gradually allow the desired loads to be absorbed. After some time, the ballasted track will meet the criteria with regard to the lateral displacement resistance. This process can last to between 0.5 and 1.5 million tonnes of traffic, but also poses the risk of irregular load application. The irregular loads arising from train traffic, e.g. due to flat spots on wheels, may result in inhomogeneous settlements.

Any necessary speed restrictions will have an impact on the capacity of the railway system and the whole network. The lower speeds will not only result in fewer trains being able to pass the section of track but may also have an impact on the timetable.

DTSs and consolidators do not necessarily have to be operated as separate machines. They are increasingly integrated into the designs of existing machines. For example, DTS and consolidators can be integrated into tamping machines or ballast distributing and profiling machines.

Network Rail has had considerable success in recent years with the use of the DTS on switch and crossing unit renewals. The particular machines used are the Unimat 09-4x4/4S Dynamic.

Fig. 5-26: Integrated consolidation unit (MATISA, left) and integrated DTS unit (Plasser & Theurer, right)

5.5 Ballast distributing and profiling machines

Before tamping, in order to have sufficient ballast available in the cribs for the formation of the bearing surface beneath each sleeper, the required quantity of ballast is deposited in the tamping area. This additional ballast will be required as a result of the correction of the track geometry and should ideally be calculated beforehand, depending on the determined correction values. Before tamping a section of track, the ballast is also deposited to the left and right of the two rails in the tamping area in order to strengthen the ballast shoulder.

Fig. 5-27: Unloading new ballast in the UK [27]

On sections of line with high speeds, it may not be permitted to pre-deposit new ballast. Due to aerodynamic effects, high train speeds may cause a suction effect under the trains [28, p. 624] that can result in ballast being whirled up. This is known as flying ballast and can severely damage technical equipment of trains and endanger lives in station areas.

For this reason, countries have adopted the approach to clear the sleeper surface of exposed ballast stones after tamping and to lower the height of the ballast crib in line with the applicable rules and standards and depending on the maximum speed of the specific section of line (see section 2.6.5). This work is usually carried out by ballast distributing and profiling machines (ballast ploughs/brushes), which will also re-establish the desired ballast profile. To do this, the machines are equipped with ballast ploughs of different profiles.

Fig. 5-28: Ballast distributing and profiling machines: R 21 (MATISA, left) and
　　　　　　USP 2010 SWS (Plasser & Theurer, right)

These ploughs make it possible to establish the special trapezoid shape of the ballast bed with the desired gradient of the ballast bed shoulder. According to EN 13231-1, the permitted maximum deviations from the specified target value are ± 10% for the gradient of the ballast bed shoulder, + 10 cm for the distance between the running edge of the rail and the ballast shoulder and + 15 % for the thickness of the ballast bed. [9]

Fig. 5-29: Different types of ploughs provide the necessary flexibility

Front, shoulder and centre ploughs (see Fig. 5-29) enable the machines to deposit and distribute the ballast. The operators of the ballast ploughs can move the ballast from the left to the right ballast shoulder or deposit it in the centre. In addition, some machines are also equipped with a ballast hopper which will enable them to pick up superfluous ballast and store it temporarily. In this way, the ballast can be picked up in a targeted manner and unloaded, if required.

Network Rail has had ballast regulators and brushes especially configured to manage the third rail in the south of England.

5.6   Summary

This chapter has looked at the correction of track geometry using on-track tamping machines. Since their introduction in the 1950s, these machines have become extremely sophisticated with advances in hydrostatics and computer technology. They are made up of thousands of parts of all kinds; mechanical, hydraulic, electric and electronic components ranging from the solid to the delicate and fragile. Together, they form a perfectly functioning machine that is supported by precise measuring and control systems and professionally trained operators.

However sophisticated and accurate the machine, it is still necessary to have sound independent acceptance and quality assurance procedures, which are the subject of the final chapter in this book.

6 Acceptance and quality assurance

Wolfgang Nemetz and Richard Spoors

6.1 Key issues

Acceptance is the declaration by customer to the contractor that the work has been carried out in accordance with the contract. This chapter deals with the acceptance procedure, the required quality certification, the underlying standards as well as the limit values applied in the DACH countries and the UK.

- What is acceptance?
- What types of acceptance are there?
- What parameters have to be observed, according to the standards, for the acceptance of track tamping work?
- How is the quality documentation with regard to track tamping work produced and which different systems are used?
- What limit values have to be adhered to, to achieve a track that will pass the acceptance criteria?
- What needs to be done if the post-measuring equipment fails?

6.2 Acceptance

The legal term acceptance is generally used for a declaration that an item or condition meets certain criteria so that work carried out will be confirmed as performance compliant. Acceptance is preceded by an inspection of the item to be accepted, where the item or its components are checked for adherence to or compliance with certain acceptance criteria. If these criteria (conditions, requirements and target results) are not met, only conditional or no acceptance will be granted, especially in the case of mandatory or other key criteria. [1] – [2]

The acceptance criteria for track are defined in European standard EN 13231-1. [3] Based on this standard, the respective railway infrastructure manager will develop relevant regulations for acceptance with detailed conditions and limit values. In addition, the respective contracts between infrastructure manager and contractor may contain supplementary requirements.

In track construction, acceptance is carried out following a quality inspection based on (largely automated) measurements which document the quality of the work. A distinction is made between preliminary and final acceptance.

6.2.1 Preliminary acceptance

In the case of maintenance and renewal work, a preliminary procedure to acceptance is usually carried out straight after the work has been completed. According to EN 13231-1, the preliminary procedure to acceptance consists of the following steps: [3]

- checking the list and quantification of the works carried out, checking the records of the measurements and tests according to the contract
- checking of documents according to the contract (e.g. drawings, notes of calculations, certificates, quality control, etc.)
- listing of works not performed although agreed in the contract
- listing of non-conformities

For a preliminary procedure to acceptance to be recognised as such, acceptance must be granted fully, with reservations or with allowances. The contractor has to be informed of the decision. If acceptance has been granted with reservations and the decision has been communicated to the contractor, the contractor will have to take the necessary actions to correct the defects within the time agreed in the contract. After that, the preliminary procedure to the acceptance has to be applied again to the corrected defects.

If the infrastructure manager decides that a defect shall not be corrected, the acceptance has to be granted with allowance according to the contract or become subject to negotiation between both parties.

6.2.2 Final acceptance

The final acceptance of the work performed must not be carried out until the track has been subjected to an appropriate passing tonnage described and defined by the infrastructure manager. The acceptance should take place within no more than six weeks or after the passage of 1.5 million load tonnes [3] after completion of the work; however, the infrastructure manager may extend the timescale to permit any follow-up tamping to be carried out.

6.2.3 Quality certification for work performed

According to EN 13231-1, the following parameters have to be checked as the basis for acceptance: [3]

- relative track geometry of plain line, switches and crossings
- absolute track position of plain line, switches and crossings (track renewals/new track construction)
- ballast profile
- dynamic stabilising work
- ballast compaction
- checking for any damage that has occurred to rails, sleepers, turnouts, fastenings, cables and other parts and for any displacement of sleepers, bearers or rail pads that occurred as a result of the work.

The infrastructure manager may request further documented measurements or tests if this has been contractually agreed. Furthermore, the selection of measuring equipment if this has been stipulated in the contract, may also be restricted.

The relative track geometry has to be measured by a track recording vehicle or by a track construction and maintenance machine fitted with measuring equipment, according to standards EN 13848-2 (track recording vehicles) [4] or EN 13848-3 (track construction and maintenance machines) [5]. If the measuring equipment fails or is not available, measurements have to be taken and documented using manual or lightweight devices, in accordance with EN 13848-4 [6]. If work on the track affects the track geometry, the relative track geometry has to be measured as per series EN 13848 before commercial trains are allowed to run on the track.

For the acceptance procedure, every section of track, every switch and crossing has to be checked by experts who have been nominated by the infrastructure manager and the contractor. The contract has to specify who carries out the measurements and how they are to be documented. These measurements are mainly referred to as post-measurements for quality control.

6.2.4 Acceptance limit values in EN 13231-1

EN 13231-1 provides acceptance limit values for tamping, lifting, lining, compacting and regulating work for the following parameters: [3]

- cross level in [mm], (deviation from design or target value)
- longitudinal level in [mm], (mean to peak)
- alignment in [mm], (mean to peak)
- twist over 3 m in [mm], (deviation from design value to peak)
- absolute track position in new track and renewals, categorisation into classes
- track gauge in [mm], (deviation from the design value)

For compliance measurements, the requirements and tolerances for the deviation from the target track geometry are used. The client or infrastructure manager must have specified that these measurements have to refer to a certain point in the track, the survey point, which defines the absolute track position. The infrastructure manager has to define the survey point and the spacing of the control measurements.

Furthermore, the infrastructure manager has to specify the external reference system, normally a network of geodetic reference points, as well as geodetic measurements for verifying the absolute track position. In addition, the absolute track position has to be measured from the geodetic reference points, e.g. for fixed point, level, satellite measurements or measurements using a tape or similar instruments.

According to EN 13231-1, it is paramount to measure and document the relative track geometry. On the other hand, there is only a recommendation for recording the following working parameters: [3]

- the lifting value for each rail at each tamping cycle
- the maximum slue value
- the tamping depth depending on the construction height of the track
- the squeezing time per squeezing operation
- the functional check of the control system before and after each tamping cycle
- working speed, vertical load and frequency for dynamic track stabilisation
- the lowering of the track from dynamic track stabilisation
- the ballast compaction working parameters (for compacting ballast shoulder and ballast between sleepers)

The infrastructure manager has to specify in the contract for each of these parameters if he accepts the work of tamping machines that are not or only partially equipped with such systems.

6.2.5 Acceptance limit values in the UK

Network Rail specifies specific track geometry parameters that must be met for new work, track renewals (capital works) and refurbishment (life extension capital works) and these are reproduced in the tables in this chapter. It also requires that the overall quality of track in each eighth of a mile shall meet target Standard Deviation (SD) values for 35 and 70 metre top and line that are banded by line speed and tabulated in the respective standards. The background to Network Rail's use of SD measurements is described in Chapter 4.

The acceptance of maintenance tamping, the majority of which is delivered through the measured compensation method, is based on the measured geometry of the completed work and assurance during the work that key requirements have been met. (see Chapter 4, section 4.5.3) Accordingly, acceptance limit values as defined in EN 13231-1 may not be specified.

6.3   Sequence of acceptance

The inspection concerning the suitability of the track to be released for operations is carried out straight after completion of the work in the case of repair and maintenance tamping operations. This implies that the infrastructure manager is taking over the work but has no significance with regard to adhering to the criteria of EN 13231-1 or the individual contract.

Once the contractor has confirmed in writing that the work has been completed (e.g. by presenting daily and weekly reports as well as the measurement sheets and the contractually agreed post-measuring documentation), the preliminary acceptance will be granted in line with the criteria listed in section 6.4. The infrastructure manager can specify additional criteria on top of the standard specifications, provided they have been agreed prior to contract conclusion.

The contractor has to be informed of the result of the preliminary procedure to acceptance. Areas which do not comply with the criteria will usually have to be revisited and any existing faults rectified.

The final acceptance is granted within a time specified by the infrastructure manager (according to the standard approx. 6 weeks or 1.5 million loaded tonnes). If the acceptance criteria are not met, the relevant provisions made in the contract will apply. With the final acceptance, any warranties from the contractor will expire, unless specified otherwise in the contract.

6.4   Acceptance of track renewals in the UK

Network Rail's acceptance of track renewals into maintenance is based primarily on compliance with the standards NR/L2/TRK/2500 – Engineering Assurance Arrangements for the Design and Construction of Track and NR/L2/TRK/2102 – Design and Construction of Track where geometric tolerances and quality parameters are defined. Major projects that include new track embrace broader acceptance requirements including the Construction (Design and Management) regulations and an Asset Management Plan concerning asset change, however, the core technical requirements for track are the same. Network Rail's acceptance process for track renewals was developed in the 1990's when Railtrack, the new Infrastructure Management company, was created (1994) and track renewals were contracted in from the former British Rail in what were known as RT1B contracts. Whilst the principles support those in EN 13231-1, it is the Network Rail standards that are followed rather than their European counterpart.

6.5   Acceptance criteria for tamping work at DB, ÖBB, SBB and NR

This section will describe the basic acceptance criteria applied in the DACH countries and the UK and will compare them with each other and with European standards.

According to EN 13231-1, the acceptance measurement of the track geometry has to proceed as follows: [3]

- The infrastructure manager has to decide if the results of the 10 m chord measurement or the D1, D2 or D3 results as per EN 13848-1 [7] will be used. D1, D2 and D3 are the three wavelength intervals in EN 13848 for the evaluation of track geometry: – D1 (3 – 25 m) – D2 (25 – 70 m) – D3 (70 – 150 m).
- The method of analysis (i.e. the target-actual comparison) must be "mean to peak".
- Maintenance machines which were supplied before the standard was published may make use of the "peak to peak" method of analysis.

- The tolerance values have to be specified by the infrastructure manager.
- Chord measurements carried out with maintenance machines have to use an asymmetric chord with a ratio of 40% to 60% and a length of 10 m. For measurements carried out with track measuring trolleys or manually operated devices, an asymmetric chord of 10 m (in curves) and 20 m (on straight track) is permitted.
- Track measuring vehicles, as well as track construction and maintenance machines that were supplied before the standard was published, are permitted to use the symmetric chord.
- If measurements are carried out using a chord measuring system with base lengths other than 10 m, the results have to be converted to 10 m of an asymmetric chord (40% to 60%).
- The twist base length normally has to be 3 m, and the analysis method shall be „zero to peak".
- In transition curves with design twist, the tolerances shall be considered from the design twist, but not from the zero line, without exceeding the intervention limit values of EN 13848-5 [8].

EN 13848-3 places certain requirements on the measuring systems of track construction and maintenance machines. [5] For example, for measuring the longitudinal level in the D1 wavelength area, the measurement must be performed using an inertial measuring system or a chord measurement based on a measuring chord with a base length of 10 m to 15 m and a chord ratio of 40 % to 60 %. The chord measurement has to undergo a recolouring process. This transforms the measurement results and equates them with the measurement values of the inertial measuring system. The alignment for the wavelength area D1, too, has to be measured on this basis.

Standard EN 13231-1 [3] stipulates the following criteria for acceptance measurements of the inner track geometry (i.e. without reference to the absolute track position):

- The measurements can be carried out using track measuring vehicles or track construction and maintenance machines. Therefore, the measurements are taken in a loaded condition.
- The distance between the points of measurement shall be no more than 0.5 m for continuous action machines and no more than 1.5 m for cyclic action machines.
- The standard also allows these acceptance measurements to be carried out by track measuring trolleys or manually operated devices.

The railway administrations of the DACH countries comply with this standard in their rules and regulations. They permit acceptance measurements to be carried out with track measuring/recording vehicles or track construction and maintenance machines (in combination with the PALAS measuring system in Switzerland). These measurements are "loaded track measurements", in line with the standard. Network Rail requires the following geometric parameters to be measured by an approved recording device following the final tamp of new work and track renewals to open the line at the permissible line speed: [9]

a) cross-level;
b) 3 m twist;
c) track gauge;
d) horizontal geometry (as horizontal versines using lining chord on tamper); and
e) vertical geometry (as vertical versines using lining chord on tamper).

Network Rail has a defects liability period following new works and track renewals during which time the track geometry quality SDs will be measured by a TRV and compared to the limit values in the Track Construction Standard [9]. The requirements following maintenance tamping include written confirmation that twist, cross-level, vertical geometry and horizontal geometry meet the track maintenance standards [10].

If the machine measuring equipment fails, the track geometry will have to be documented with manual post-measurements. For this, the distance between the points of measurement should be 1 m in Germany and a minimum of 5 m in Austria. In Germany and Austria, this manual measurement may only be used for commissioning but not for acceptance.

The measurement parameters are similar in the countries; however, some details with regard to the measurement base and the methods of analysis are different. Amongst other things, this is due to the different post-measuring systems (MCRS Multi-Channel Recording System or DRS Digital Recording System, see section 6.6).

According to EN 13231-1, the **track gauge** has to be presented as a deviation from the target value (target-actual comparison). [3] Furthermore, the measurement values of the **cross level** have to be presented as a deviation from the target value (peak to peak). This is common practice at all DACH railways and in the UK.

For measuring the **longitudinal level** (base length 10 m), the standard defines that the measurement values are presented as the difference of the mean value to the peak value (mean-peak comparison).

When using the MCRS system for measuring the longitudinal level, DB stipulates the analysis method from the peak of the measurement point to the peak of the next measurement point. The reason for this is that the MCRS is technically not equipped to calculate a mean value. When using a DRS recorder, the analysis of the measurements requires the comparison of the zero value to the peak measurement value (zero-peak comparison).

ÖBB stipulates that the longitudinal level is measured as the difference between the mean value (for a base length of 20 m) and the peak value (mean-peak comparison).

SBB uses the analysis method of the zero value to the peak value for chord measurements (zero-peak comparison).

For measuring the **longitudinal level based on wavelengths D1, D2 and D3**, the standard specifies the mean value to peak value approach to be used for analysis (mean-peak comparison). With this approach, only SBB has so far introduced limit values for the wavelengths D1 and D2 (mean-peak comparison). As with the longitudinal level, the standard defines that the **alignment** (base length 10 m) is presented as the difference of the mean to the peak measurement value (mean-peak comparison).

When using the MCRS system for measuring the alignment, Germany stipulates the analysis method from the peak of the measurement point to the peak of the next measurement point (peak-peak comparison). Here, too, the MCRS is not technically equipped to calculate a mean value. When using a DRS recorder, the analysis of the measurements requires the comparison of the zero to the peak measurement value (zero-peak comparison).

Austria stipulates that the alignment measurement is presented as the difference of the target value to the actual measurement value (zero-peak comparison). In addition, adjacent versine differences are compared (at 5 m intervals). These difference-from-difference comparison values (maximum deviation of the difference) also have limit values.

Switzerland uses the analysis method of the zero to the peak measurement value for chord measurements (zero-peak comparison).

For measuring the **alignment based on wavelengths D1, D2 and D3**, the standard specifies the mean value to peak value approach to be used for analysis (mean-peak comparison), as with the longitudinal level measurement. With this approach, only SBB has so far introduced limit values for the wavelengths D1 and D2 (mean-peak comparison).

The evaluation method for **twist** is defined in the standard by the deviation from the target value at a base length of 3 m (target-peak comparison).

In Austria and Switzerland, the deviations from the target value at a base length of 3 m are analysed. In Germany, measuring the twist is not explicitly required for acceptance; therefore, no limit values are defined.

Acceptance limit values for track gauge for new track, renewals and maintenance:

	$v \leq 40$ km/h	$v \leq 80$ km/h	80 km/h $< v \leq$ 120 km/h	120 km/h $< v \leq$ 160 km/h	160 km/h $< v \leq$ 230 km/h	230 km/h $< v \leq$ 360 km/h
EN 13231-1	+ 4/- 3 (+ 7/- 3)		+ 4/- 3 (+ 5/- 3)	+ 4/- 3 (+ 5/- 3)	+ 4/- 2 (+ 5/- 2)	+ 3/- 2 (+ 4/- 2)
DB	+ 4/- 2 (N/A)		+ 4/- 2 (N/A)		+ 3/- 2 (N/A)	
ÖBB	+ 3/- 3 (+ 8/- 3)	+ 3/- 3 (+ 5/- 3)	+ 2/- 2 (+ 5/- 3)	+ 2/- 2 (+ 4/- 3)	+ 2/- 2 (+ 3/- 3)	
SBB	+ 3/- 3 (+ 3/- 3)		+ 4/- 2 (+ 4/- 2)	+ 3/- 2 (+ 3/- 2)	+ 3/- 1 (+ 3/- 1)	

NR	$v \leq 30$ km/h	35 km/h $\leq v \leq$ 65 km/h	70 km/h $\leq v \leq$ 100 km/h	105 km/h $\leq v \leq$ 155 km/h	160 km/h $\leq v \leq$ 200 km/h	> 200 km/h
	+ 6/0	+ 6/0	+ 6/0	+ 5/0	+ 5/0	+ 5/0
	($v \leq 30$ km/h)	(35 km/h $\leq v \leq 95$ km/h)		(100 km/h $\leq v \leq$ 145 km/h)	(> 150 km/h)	
	(+30/-9)	(+20/-9)		(+15/-6)	(+15/-5)	

The main values apply to new track and track renewal, the values in brackets apply to maintenance.
N/A: No values provided or specified.

Fig. 6-1: Acceptance limit values for track gauge in mm [11] [12]

The acceptance limit values in the respective countries are divided into different speed categories, in line with EN 13231-1. [3] Special features at ÖBB and Network Rail are the use of limit values at the slower speeds of ≤ 40 and 30 km/h respectively.

The following example illustrates this for ÖBB:

A section of track is renewed in the ÖBB network. The maximum permitted line speed is 150 km/h. The acceptance limit values applied are those for new construction and renewal in the speed category higher than 120 km/h up to and including 160 km/h; these are defined with a tolerance range of +2 to –2 mm.

Acceptance limit values for cross level for new track, renewal and maintenance:

	v ≤ 40 km/h	v ≤ 80 km/h	80 km/h < v ≤ 120 km/h	120 km/h < v ≤ 160 km/h	160 km/h < v ≤ 230 km/h	230 km/h < v ≤ 360 km/h
EN 13231-1	± 3 (± 5)		± 3 (± 4)	± 3 (± 4)	± 2 (± 3)	± 2 (± 3)
DB	± 3 (± 5)		± 3 (± 4)		± 2 (± 3)	
ÖBB	± 3 (± 6)	± 3 (± 5)	± 3 (± 4)	± 2 (± 3)	± 2 (± 3)	
SBB	± 3 (± 4)		± 3 (± 4)	± 2 (± 3)	± 2 (± 3)	
NR	v ≤ 30 km/h	v 35 km/h ≤ v ≤ 65 km/h	v 70 km/h ≤ v ≤ 100 km/h	105 km/h ≤ v ≤ 155 km/h	160 km/h ≤ v ≤ 200 km/h	> 200 km/h
	±/-10 (±/-20)	±/-8 (±/-20)	±/-8 (±/-20)	±/-5 (±/-15)	±/-3 (±/-15)	±/-2

The main values apply to new track and track renewal, the values in brackets apply to maintenance.

Fig. 6-2: Acceptance limit values for cross level in mm [11] [12]

The acceptance limits for cross level, as defined in the standard, have largely been adopted by the individual railway administrations and even been tightened in some cases (e.g. DB, ÖBB and SBB in the speed category 120 km/h < v ≤ 160 km/h).

Acceptance limit values for longitudinal level using chord measurement (10 m) for new track, renewal and maintenance:

	Parameter	v ≤ 40 km/h	v ≤ 80 km/h	80 km/h < v ≤ 120 km/h	120 km/h < v ≤ 160 km/h	160 km/h < v ≤ 230 km/h	230 km/h < v ≤ 360 km/h
EN 13231-1	Longitudinal level [mm]	± 6 (± 7)		± 4 (± 5)	± 4 (± 5)	± 3 (± 4)	± 3 (± 4)
DB	Longitudinal level (peak to peak) [mm]	± 5 (± 6)		± 4 (± 5)		± 3 (± 4)	
DB	Longitudinal level (zero to peak) [mm]	± 4 (± 5)		± 3 (± 4)		± 2 (± 3)	
ÖBB	Longitudinal level [mm]	± 7 (± 8)	± 6 (± 7)	± 4 (± 5)	± 4 (± 5)	± 3 (± 4)	N/A (N/A)
SBB	Longitudinal level (zero to peak) [mm]	± 5 (± 5)		± 4 (± 4)	± 3 (± 3)	± 3 (± 3)	N/A (N/A)
NR*		v ≤ 30 km/h	v 35 km/h ≤ v ≤ 65 km/h	v 70 km/h ≤ v ≤ 100 km/h	105 km/h ≤ v ≤ 155 km/h	160 km/h ≤ v ≤ 200 km/h	> 200 km/h
	Permissible rate of change from design between adjacent 10 m offsets [mm] **in reverse curves**	15 (25)	10 (20)	8 (16)	6 (12)	5 (10)	2 (N/A)
	Permissible rate of change from design between adjacent 10 m offsets [mm] **elsewhere**	25 (25)	15 (20)	13 (16)	10 (12)	7 (10)	3 (N/A)

The main values apply to new track and track renewal, the values in brackets apply to maintenance.
N/A: No values provided or specified.
* Network Rail uses a different measure definition to that specified in EN13231-1.

Fig. 6-3 : Acceptance limit values for longitudinal level – 10 m chord measurement mean to peak [11] [12]

Fig. 6-3 shows the permitted acceptance limit values for the longitudinal level on the basis of a chord measurement. The measurement base here is a chord measured off-centre that is then converted to a chord length of 10 m.

The railway administration can decide in favour of this measurement method or the one shown in Fig. 6-4 (measuring longitudinal level based on wavelengths D1, D2 and D3 using the inertial system).

Network Rail uses a slightly different measurement criterion such that their values are not directly comparable.

Acceptance limit values for longitudinal level based on wavelengths D1, D2 and D3 for new track, renewal and maintenance:

	Wavelength area	v ≤ 80 km/h	80 km/h < v ≤ 120 km/h	120 km/h < v ≤ 160 km/h	160 km/h < v ≤ 230 km/h	230 km/h < v ≤ 360 km/h
EN 13231-1	D1	± 4 (± 5)	± 3 (± 4)	± 3 (± 4)	± 2 (± 3)	± 2 (± 3)
	D2		N/A (N/A)		± 3 (± 4)	± 2 (± 3)
	D3		N/A (N/A)			planned
SBB	D1	± 2.5 (± 2.5)	± 2.0 (± 2.0)	± 1.5 (± 1.5)	± 1.5 (± 1.5)	N/A (N/A)
	D2		N/A (N/A)		± 2.0 (± 2.0)	N/A (N/A)
	D3		N/A (N/A)			

The main values apply to new track and track renewal, the values in brackets apply to maintenance.
DB, NR and ÖBB have not specified any values for this parameter.
N/A: No values provided or specified.

Fig. 6-4: Acceptance limit values for the longitudinal level of D1, D2 and D3 in mm

Fig. 6-4 shows the permitted acceptance limit values of the longitudinal level based on wavelengths D1, D2 and D3. This measurement is usually carried out using inertial measuring systems. The railway administration can decide in favour of this type of measurement or the one shown in Fig. 6-3 (measuring longitudinal level using chord measurement).

6 Acceptance and quality assurance

Acceptance limit values for alignment using chord measurement (10 m) for new track, renewal and maintenance:

	Parameter	v ≤ 40 km/h	v ≤ 80 km/h	80 km/h < v ≤ 120 km/h	120 km/h < v ≤ 160 km/h	160 km/h < v ≤ 230 km/h	230 km/h < v ≤ 360 km/h
EN 13231-1	Alignment [mm]	± 5 (± 7)		± 3 (± 5)	± 3 (± 5)	± 3 (± 4)	± 3 (± 4)
DB	Alignment (peak to peak) Measurement base 4.0/6.0 m [mm]	± 5 (± 6)		± 4 (± 5)	± 3 (± 4)		N/A (N/A)
DB	Alignment (zero to peak) Measurement base 4.0/6.0 m [mm]	± 4 (± 5)		± 3 (± 4)	± 2 (± 3)		N/A (N/A)
ÖBB	Alignment target/actual (zero to peak) [mm]	± 6 (± 8)	± 5 (± 7)	±4 (± 5)	± 3 (± 5)	± 3 (± 4)	N/A (N/A)
ÖBB	Maximum deviation of difference [mm]	7 (7)	5 (6)	4 (5)	3 (5)	3 (4)	N/A (N/A)
SBB	Alignment (zero to peak) [mm]	± 5 (± 5)		± 4 (± 4)	± 3 (± 3)	± 3 (± 3)	N/A (N/A)

		v ≤ 30 km/h	v 35 km/h ≤ v ≤ 65 km/h	v 70 km/h ≤ v ≤ 100 km/h	105 km/h ≤ v ≤ 155 km/h	160 km/h ≤ v ≤ 200 km/h	> 200 km/h
NR*	Permissible rate of change from design. Either the difference between adjacent 10 m offsets or the difference between consecutive offsets on overlapping 20 m chords [mm] (Maximum permitted horizontal measurement between adjacent 10 m offsets)	15 (25)	10 (20)	8 (16)	6 (12)	5 (10)	3 (N/A)

The main values apply to new track and track renewal, the values in brackets apply to maintenance.
N/A: No values provided or specified.
* Network Rail uses a different measure definition to that specified in EN13231-1

Fig. 6-5: Acceptance limit values for alignment – 10 m chord measurement [11] [12]

Fig. 6-5 shows the permitted acceptance limit values for alignment, using an off-centre chord measurement that is converted to a 10 m base length. The railway administration can decide in favour of this measurement method or the one shown in Fig. 6-6 (measuring alignment based on wavelengths D1, D2 and D3 using the inertial system).

Network Rail uses a slightly different measurement criterion such that their values are not directly comparable

Acceptance limit values for alignment based on wavelengths D1, D2 and D3 for new track, renewal and maintenance:

	Wavelength area	v ≤ 40 km/h	v ≤ 80 km/h	80 km/h < v ≤ 120 km/h	120 km/h < v ≤ 160 km/h	160 km/h < v ≤ 230 km/h	230 km/h < v ≤ 360 km/h
EN 13231-1	D1	± 4 (± 5)	± 2 (± 4)	± 2 (± 4)	± 2 (± 3)	± 2 (± 3)	± 2 (± 3)
	D2	N/A (N/A)			± 3 (± 4)	± 2 (± 3)	± 2 (± 3)
	D3	N/A (N/A)				planned	planned
SBB	D1	± 2.5 (± 2.5)		± 2.0 (± 2.0)	± 1.5 (± 1.5)	± 1.5 (± 1.5)	N/A (N/A)
	D2	N/A (N/A)				± 2.0 (± 2.0)	N/A (N/A)
	D3	N/A (N/A)					

The main values apply to new track and track renewal, the values in brackets apply to maintenance. DB, NR and ÖBB have not specified any values for this parameter.
N/A: No values provided or specified.

Fig. 6-6: Acceptance limit values for the alignment of D1, D2 and D3 in mm

Fig. 6-6 shows the permitted acceptance limit values of the alignment based on wavelengths D1, D2 and D3. This measurement is usually carried out using inertial measuring systems. The railway administration can decide in favour of this type of measurement or the one shown in Fig. 6-5 (measuring alignment using chord measurement).

Acceptance limit values for twist for new track, renewal and maintenance:

	Parameter	v ≤ 40 km/h	v ≤ 80 km/h	80 km/h < v ≤ 120 km/h	120 km/h < v ≤ 160 km/h	160 km/h < v ≤ 230 km/h	230 km/h < v ≤ 360 km/h
EN 13231-1	Twist over 3 m [mm]	± 4.5 (± 4.5)		± 3 (± 4.5)	± 3 (± 4.5)	± 3 (± 3)	± 3 (± 3)
DB	Twist acceptance value ensured through cross level						
ÖBB	Twist (target/actual) [mm]	± 4.5 (± 5)	± 4.5 (± 4.5)	± 3 (± 3)	± 3 (± 3)	± 2 (± 2)	N/A (N/A)
	Maximum deviation from specified value [‰]	± 0.75 ‰ (± 0.75 ‰)		± 0.75 ‰ (± 0.75 ‰)	± 0.5 ‰ (± 0.5 ‰)	± 0.5 ‰ (± 0.5 ‰)	N/A (N/A)
SBB	Maximum value [‰]	2.5 ‰ (2.5 ‰)		2.5 ‰ (2.5 ‰)	2 ‰ (2 ‰)	1 ‰ (1 ‰)	N/A (N/A)

	Parameter	v ≤ 30 km/h	v 35 km/h ≤ v ≤ 65 km/h	v 70 km/h ≤ v ≤ 100 km/h	105 km/h ≤ v ≤ 155 km/h	160 km/h ≤ v ≤ 200 km/h	> 200 km/h
NR*	Transition Curve [mm]	8	8	8	7	6	6
	Elsewhere [mm]	5	5	5	5	3	3

The main values apply to new track and track renewal, the values in brackets apply to maintenance.
N/A: No values provided or specified.
* Twist faults (measured over 3 m) worse than 1 in 200 shall not be permitted to remain in the track. When twist faults are discovered they shall be repaired within a timescale commensurate with the risk of derailment. For maintenance tamping the limit value is 1/300.

Fig. 6-7: Acceptance limit values for twist [10] [11] [13]

Fig. 6-7 shows the permitted acceptance limit values for twist. These limit values refer to a measurement base length of 3 m.

Unlike ÖBB, SBB and NR, DB does not define any specific limit values for twist. This is based on the assumption that the adherence to the acceptance limit values for the cross level ensures that the twist tolerances cannot be exceeded.

The different methods for including the **outer (absolute) track position** in the acceptance measurements are described below. Standard EN 13231-1 stipulates that the requirements and tolerances for a new track and track renewals (including switches and crossings) are applied here [3], whereas these are voluntary for the infrastructure manager for the acceptance following maintenance work.

At DB, the minimum required used to be the checking of the geodetic position (level and distance) as a target-actual comparison referenced to the route layout plan. As of 01.01.2020, a geodetic acceptance measurement is required after any changes to tracks and turnouts as well as constraint points in the route layout, if the changes affect the position of these installations.

ÖBB has defined tolerance ranges for the longitudinal level that differentiate between renewal/new construction and maintenance work. The tolerance ranges for alignment apply throughout.

SBB specifies permitted maximum deviations from the reference or fixed point (e.g. measuring rod or measuring bolt on the catenary support). By using Toporail and free stationing (= coordinative measurement), the absolute target and actual track geometry can be defined and measured at any point.

The majority of Network Rail's track falls under the definition in EN 13231-1 as Relative Track Geometry. The fast lines of the West Coast Main Line from London to Glasgow are the exception and are managed to the definition of Absolute Track Position, which includes the requirements of relative track geometry.[14] All track in new works and track renewals will have their as built position measured in absolute terms, by reference to external references or fixed points.

The following table shows the permitted tolerance values.

Acceptance limit values for the outer (absolute) track position:

		Vertical position [mm]	Horizontal position [mm]	Remarks
EN 13231-1	AP1	± 10	± 10	An accepted track has to correspond with the selected categories (AP 1 to AP 4) of EN 13231-1.
	AP2	± 15	± 15	
	AP3	± 20	± 20	
	AP4	± 25	± 25	
DB		Since 2018 should, from 01.01.2020 must be: -25 ≤ TG ≤ + 15	Since 2018 should, from 01.01.2020 must be: -15 ≤ TG ≤ +15	Geodetic acceptance tolerance (TG) = Implementation tolerance (TA) + Measuring tolerance (TM)
ÖBB		± 10/- 20		Maintenance
		± 10/- 20		Renewal and new track: modified target level of - 10 mm. Tolerances for acceptance refer to this modified target level.
			± 10	Turnouts and track up to R ≥ 300 m
			± 15	Track R ≥ 300 m

SBB	Parameter	v ≤ 80 km/h	80 km/h < v ≤ 120 km/h	120 km/h < v ≤ 160 km/h	160 km/h < v ≤ 200 km/h	Remarks
	Absolute alignment	± 5	± 5	± 4	± 4	Applies to 3rd tamping cycle and post-tamping
		± 6	± 6	± 4	± 4	Applies to maintenance tamping
	Absolute longitudinal level	± 10				3rd tamping cycle: modified target geometry = -10 mm; Post-tamping: target geometry = 0 mm

NR	Parameter	v ≤ 35 km/h	v 40 ≤ v ≤ 65 km/h	v 70 ≤ v ≤ 95 km/h	100 km/h ≤ v ≤ 155 km/h	160 km/h ≤ v ≤ 200 km/h	> 200 km/h	Remarks
	Vertical [mm]	+10/-20	+10/-20	+10/-20	+10/-20	+10/-20	+10/-20	New Work
	Horizontal [mm]	±15	±15	±15	±15	±10	±10	
	Vertical [mm]	±25	±25	±25	±25	±25	(N/A)	Maintenance
	Horizontal [mm]	+10/-20	+10/-20	+10/-20	+10/-20	+10/-20	(N/A)	
	Full Tolerance H [mm]	±25	±25	±25	±25	±25	(N/A)	Maintenance of track under overhead line electrification, horizontal (H) and vertical (V)
	Full Tolerance V [mm]	±25	±25	±25	±25	±25	(N/A)	
	Close Tolerance H [mm]	±15	±15	±15	±15	±15	(N/A)	
	Close Tolerance V [mm]	±15	±15	±15	±15	±15	(N/A)	

Fig. 6-8: Acceptance limit values for absolute track position [10] [12] in mm

According to EN 13231-1, the customer has to categorise the track into classes from AP 1 to AP 4. [3] This classification determines the requirements for position accuracy of the absolute track position.

6.6 Machine-integrated recording systems

According to EN 13231-1 [3], track and switch tamping machines have to be equipped with measuring systems as per EN 13848-3 [5] for measuring the inner track geometry. The measuring systems are integrated into the tamping machines and document the final position of the track during the work process.

The post-measuring system has to be independent from the measuring systems for machine control. This measuring system records the tamped and stabilised track geometry mechanically in the track area that is no longer affected by the machine work procedure. The measurement values are converted into electric signals and output to the recording devices.

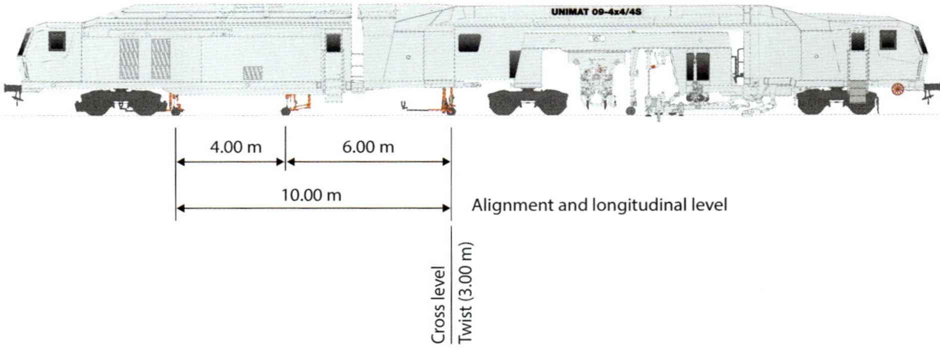

Fig. 6-9: Example of a post-measuring system

To compensate for the different construction lengths of the machines and thus different chord lengths, the signals have to be converted to a standardised measurement base of 10 m in the off-centre division ratio of 4:6.

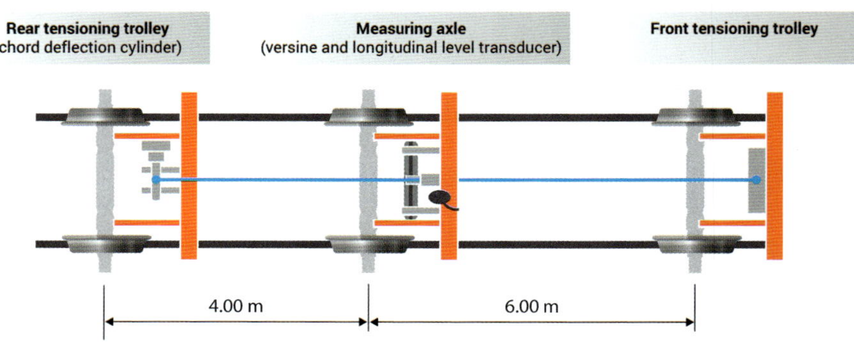

Fig. 6-10: Measuring axles of the post-measuring system

As these are primarily mechanical transducer systems, regular maintenance of the measuring equipment is crucial for the reliability of the system. The post-measuring systems have to be approved before their first use and then checked and released for use by the infrastructure manager at regular intervals. In addition, every time the system is deployed, a self-test has to be carried out by triggering a test deflection; this has to be documented (before or before and after the measurement).

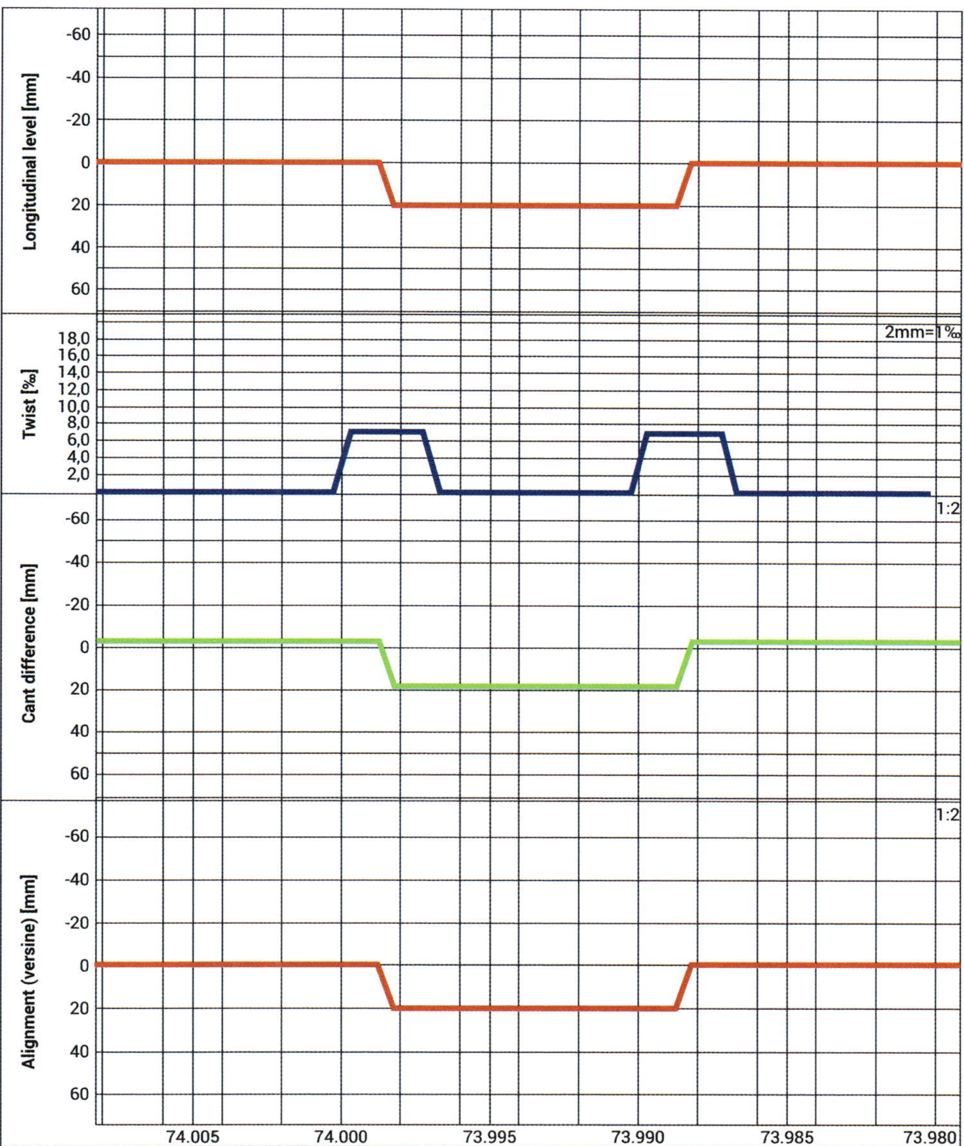

Fig. 6-11: Test deflection of a multi-channel recording system (MCRS)

Here, each channel of the recording system has to deflect by a defined value and then return to the starting position.

If the test deflections of the individual channels do not conform with the specifications, the measurement will not be accepted.

Measuring has to start at least 20 m before the tamping area and finish at least 20 m after it to reflect the transition from and to the non-tamped track section.

Different types of recording systems are available. The multi-channel recorder with 6 channels (MCRS-6) is an industrial mechanical recording system. The recorded measurement values of the transducers are electronically converted in a (sealed) switch cabinet and output directly to the recorder. Depending on their design, MCRS recorders can output two, three or six measuring channels.

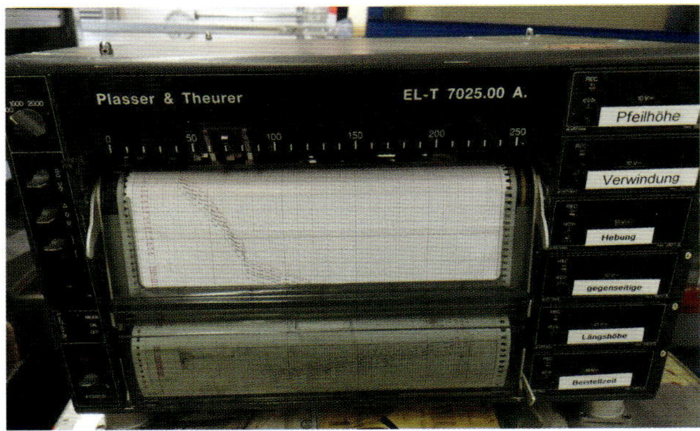

Fig. 6-12: 6-channel MCRS recorder

The DAR recorder is a development of the MCRS-6 recording system. This is a multi-channel recorder with 8 channels (MCRS-8) called data acquisition recorder (DAR). In addition to the mechanical recording system, the DAR has a computing unit that enables data recording and data storage.

The DAR can take over the basic data of the post-measurement documentation from the control computer of the machine and generate calculations of the measurement data. In addition, the measurement data can be output on data media. The DAR has the capacity to output up to eight channels mechanically on paper and digitally on the data medium.

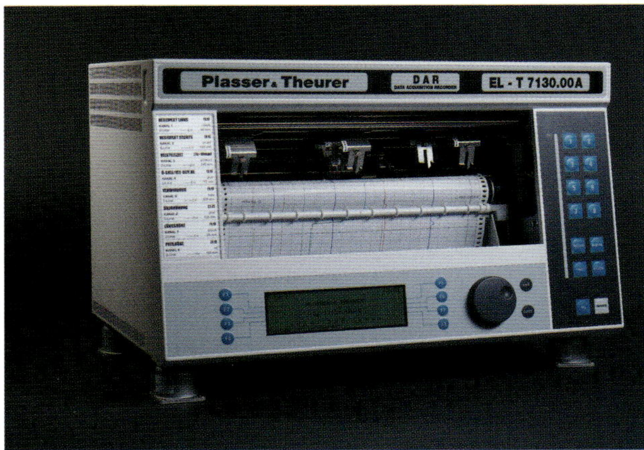

Fig. 6-13: 8-channel data acquisition recorder (DAR)

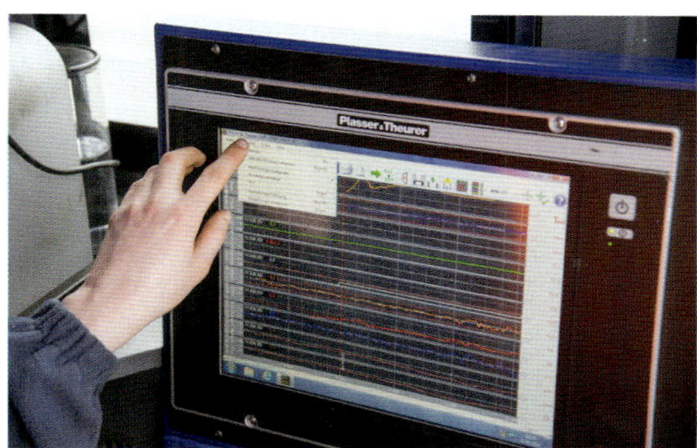

Fig. 6-14: Touch screen of the DRP

The DRP (Data Recording Processor) is a digital recording system (DRS) that processes the measurement data digitally, visualises it on a screen in real time, compares it to the limit values and stores it. The individual measurement parameters are provided as analogue signals by the measuring transducers and recorded and stored by the DRP system in digital format at regular intervals (e.g. every 25 cm). The measured values can be displayed on the touch screen in real time and, if required, printed out on commercially available printers.

The comparison of the measured values with the specified, railway-specific limit values of the individual parameters makes it possible for exceeded tolerances to be identified during the work process. The exceeded values are automatically displayed graphically and listed in the exceedance report.

The DRP is capable of processing more than the previously common eight parameters. The measuring transducers of the track maintenance machines remain unchanged compared to the use of electro-mechanical recording systems, e.g. the DAR multi-channel recorder.

An automatic test deflection of the measuring transducers is performed before the start of each work procedure. The measured deflections are stored and automatically checked. The test result is also documented.

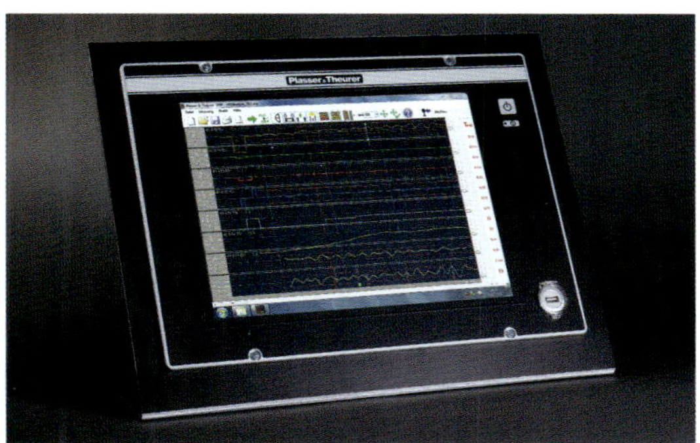

Fig. 6-15: Digital Recording Processor (DRP)

The exceedance reports provide a quick overview of standard deviations and exceeded tolerances and thus also of the quality of the track condition. Standard deviations are calculated with reference to alignment and level values for the given wavelength areas (D1: 3 – 25 m).

As an option and depending on the specifications of the railway companies, the DRP also calculates quality parameters and displays them in colour bar graphs. The colour bars divide the recording into 200-m sections and provide a quick overview of the quality areas of the worked track. Alignment and level defects can be calculated for the wavelength areas D1 - D3, in line with EN 13848-3.

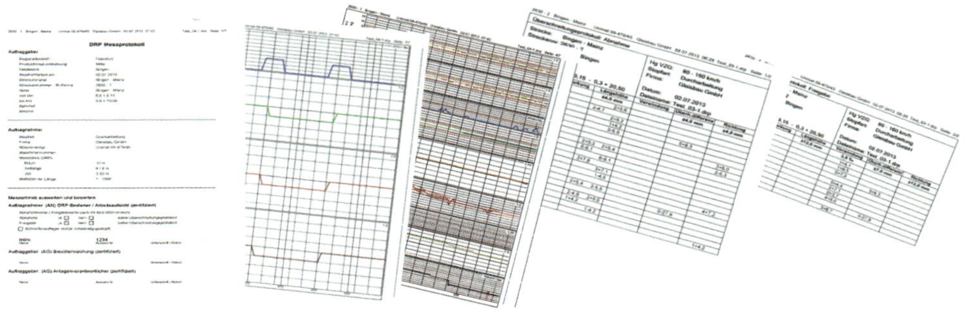

Fig. 6-16: DRS logs: cover page, test deflection, measurement records and reports

The DACH countries have different requirements of the recording systems. An overview of the required characteristics and settings of the systems in the DACH countries is given below.

6-channel MCRS					8-channel MCRS				
Position	Parameter	Colour	Scale	Remark	Position	Parameter	Colour	Scale	Remark
1	Longitudinal level	brown	1 : 1	Left rail	1	Longitudinal level	brown	1 : 1	Left rail
2*	Lifting value left rail	purple	1 : 2	Left rail*	2	Lifting value left rail	purple	1 : 2	Left rail*
	Cant difference	blue	2 mm = 1 ‰	Base: 3 m	3	Cant difference	green	1 : 1	
2*	Lifting value right rail	purple	1 : 2		4	Lifting value right rail	purple	1 : 2	
3	Cross level	green	1 : 2		5	Cross level	green	1 : 2	
4	Alignment	red	1 : 1		6	Alignment	red	1 : 1	
5	Twist	blue	1 : 1	3 mm = 1 ‰	7	Twist	blue	1 : 1	3 mm = 1 ‰
6	Squeezing time	black	10 mm = 1 s		8	Squeezing time	black	10 mm = 1 s	

Fig. 6-17: DB requirements of MCRS measurements in working mode

Position	Parameter	Colour	Scale	Remark
1	Longitudinal level	brown	1 : 2	Left rail
2	Longitudinal level	brown	1 : 2	Right rail
3	Twist	blue	2 mm = 1 ‰	Base: 3 m
4	Cant difference (target-actual)	green	1 : 2	
5	Cant	green	1 : 10	
6	Alignment	red	1 : 2	
7	Penetration depth	black	1 : 30	
8	Maximum lift	black	1 : 5	
9	Squeezing time	black	10 mm = 1 s	
10	Marking line	black	1 : 1000	

Fig. 6-18: DB requirements of digital recording systems for working mode (tamping and measuring)

Position	Parameter	Colour	Scale	Remark
1	Longitudinal level	brown	1 : 2	Left rail
2	Longitudinal level	brown	1 : 2	Right rail
3	Twist	blue	2 mm = 1 ‰	Basis: 3 m
4	Cant difference (target-actual)	green	1 : 2	
5	Cant	green	1 : 10	
6	Alignment	red	1 : 2	
7	Speed	black	2 mm = 1 km/h	Maximum v = 3.5 km/h
8	Marking line	black	1 : 1000	

Fig. 6-19: DB requirements of digital recording systems for measuring run recordings (separate measuring run)

ÖBB stipulates the use of a digital recording system. Multi-channel recorders are not permissible.

Position	Parameter	Colour	Scale	Remark
1	Longitudinal level	brown	1 : 2	Including tolerance range
2	Lift	orange	1 : 4	
3	Cant difference	purple	2 : 1	Including tolerance range
4	Versine difference	magenta	2 : 1	Including tolerance range
5	Measured cant	green	1 : 5	
6	Measured versine	red	1 : 2	
7	Twist	blue	2 : 1	Including tolerance range
8	Squeezing time	black	10 : 1	10 mm = 1.2 s

The contractor will receive the specified values for the basic data electronically via the iGleis database.
The electronic data must be output in .obbdrp data format.
Additional measuring channels are only displayed electronically, but not in the printout.

Fig. 6-20: ÖBB requirements of DRS measurements

Parameter	Scale	Remark
Curve of reference rail	1 : 1	Measurement base: 10 m
Cant	1 : 2	
Twist	1 : 1	Measurement base: 3 m
Longitudinal profile of track axis	1 : 1	Referenced to track axis

Fig. 6-21: SBB requirements of recording systems for tamping work

If a post-measuring system fails (this also includes a failed test deflection), all DACH countries require in principle for the work to be stopped. An alternative measurement (e.g. manually or with the use of an alternative measuring system) has to document the track sections that have already been worked on.

Certain conditions apply to the reading and interpretation of records from machine-integrated post-measurements. While digital recording systems supply an exceedance report, in addition to the measurement records, which outputs any exceeded tolerances and their positions, measurements taken with MCRS systems have to be checked manually. DB Netz AG has developed templates for this; they allow the user to recognise any deviations and exceeded tolerance limits in MCRS recording systems more efficiently.

ÖBB uses the hard copy of the measurement records and the report only for opening the line to traffic. For acceptance, the DRS post-measurement data is input electronically into the central track database (iGleis). There, the post-measurement values recorded are compared electronically to the target values. The result is output in a separate report and also archived in the database in a legally secure manner.

SBB checks the measurement records of the machine post-measurement and documents the results for installation checks on the respective form. Separate forms are available for the first and second tamping cycle. The third tamping cycle, the initial tamping cycle (designated "R1") and maintenance tamping are documented on another form.

Network Rail does not specifically mandate the use of independent geometry data recording devices such as DRP where these have been fitted to modern tamping machines. [15] Instead, the site engineer has the option to use the machine's guiding computer in a post tamping recording run. Time must be allowed in the work possession for the machine to run over the newly tamped track and record the geometry. This is then used to provide quality assurance reports that the new work falls within the line speed acceptance criteria for cross level, twist, gauge, top and line before the track can be returned to traffic.

Bibliography

Bibliography – Chapter 1

[1]   German Imperial Court: Betrieb einer Eisenbahn und Betriebsunternehmen im Sinne des Reichshaftpflichtgesetzes, 17.03.1879. https://opinioiuris.de/entscheidung/1724, accessed on 26.07.2018.

[2]   Directive 2001/12/EC of the European Parliament and of the Council of 26 February 2001 amending Council Directive 91/440/EEC on the development of the Community's railways. Official Journal of the European Communities, 15.03.2001. https://eur-lex.europa.eu/legal-content/EN/TXT/PDF/?uri=CELEX:32001L0012&from=EN, accessed on 10.01.2019.

[3]   Endlicher, K.-O. and Rischanek, A.: Normen im Eisenbahnwesen – Der praktische Gesamtüberblick. Austrian Standards plus Publishing, 2017.

[4]   EN 45020: Standardisation and related activities – General vocabulary, August 2007. British Standards Institution.

[5]   1846 act of parliament: The Railway – British track since 1804 by Andrew Dow, published by Pen and Sword books, Barnsley, England.

[6]   Federal Republic of Germany: Allgemeines Eisenbahngesetz (AEG). Federal Ministry of Justice and Consumer Protection, 27.12.1993, version of 29.11.2018. https://www.gesetze-im-internet.de/aeg_1994/AEG.pdf, accessed on 11.01.2019.

[7]   Federal Republic of Germany: Gesetz über die Eisenbahnverkehrsverwaltung des Bundes (Bundeseisenbahnverkehrsverwaltungsgesetz – BEVVG). Federal Ministry of Justice and Consumer Protection, 27.12.1993, version of 29.11.2018. https://www.gesetze-im-internet.de/aeg_1994/AEG.pdf, accessed on 10.01.2019.

[8]   Federal Republic of Germany: Eisenbahn-Bau- und Betriebsordnung (EBO). Federal Ministry of Justice and Consumer Protection, 08.05.1967, version of 26.07.2017. https://www.gesetze-im-internet.de/ebo/EBO.pdf, accessed on 11.01.2019.

[9]   Federal Railway Office: Eisenbahnspezifische Liste Technischer Baubestimmungen (ELTB), 18.01.2016. https://www.eba.bund.de/SharedDocs/Downloads/DE/ Infrastruktur/AllgemeineVorschriften/ELTB/21-ELTB_01_2016_Aenderungen.pdf?__blob=publicationFile&v=2, accessed on 12.07.2018.

[10]   Freystein, H., Muncke, M., Schollmeier, P.: Handbuch Entwerfen von Bahnanlagen – Regelwerke, Planfeststellung, Bau, Betrieb, Instandhaltung. PMC Media House GmbH, Leverkusen, 2015.

[11]   Republic of Austria: Bundesgesetz über Eisenbahnen (EisbG). Federal Ministry for Digital and Economic Affairs, 07.03.1957, version of 14.01.2019. https://www.ris.bka.gv.at/GeltendeFassung.wxe?Abfrage=Bundesnormen&Gesetzesnummer=10011302, accessed on 11.01.2019.

[12]   Commission Regulation (EU) No 1299/2014 of 18 November 2014 on the technical specifications for interoperability relating to the 'infrastructure' subsystem of the rail system in the European Union. Official Journal of the European Union, 12.12.2014. http://eur-lex.europa.eu/legal-content/EN/TXT/PDF/?uri=CELEX:32014R1299&from=EN, accessed on 11.01.2019.

[13]   EN 13848-5: Railway applications – Track – Track geometry quality – Part 5: Geometric quality levels – Plain line, switches and crossings, September 2017. British Standards Institution.

[14]   Republic of Austria: Verordnung über den Bau und Betrieb von Eisenbahnen (Eisenbahnbau- und -betriebsverordnung – EisbBBV). Federal Ministry for Digital and Economic Affairs, 19.11.2008, version of 26.06.2014. https:// www.ris.bka.gv.at/GeltendeFassung.wxe?Abfrage=Bundesnormen&Gesetze snummer=20006077, accessed on 11.01.2019.

[15]   Swiss Confederation: Eisenbahngesetz (EBG). Federal Office, 20.12.1957, version of 01.01.2018. https://www.admin.ch/opc/de/classified-compilati on/19570252/2 01801010000/742.101.pdf, accessed on 16.07.2018.

[16]   Swiss Federal Council: Verordnung über Bau und Betrieb der Eisenbahnen (EBV). Federal Office, 23.11.1983, version of 15.05.2018. https://www. admin.ch/opc/de/classified-compil ation/19830331/201805150000/742.141.1.pdf, accessed on 11.01.2019.

[17]   Swiss Department of Environment: Ausführungsbestimmungen zur Eisenbahnverordnung (AB-EBV). Federal Office, 01.06.2016. https://www.bav.admin.ch/ bav/de/home/ rechtliches/rechtsgrundlagen-vorschriften/ab-ebv.html, accessed on 16.07.2018.

[18]   https://www.railwaysarchive.co.uk/docsummary.php?docID=57, accessed 07/09/2020

[19]   Hall, Stanley (1990). Railway Detectives: The 150-year Saga of the Railway Inspectorate. Shepperton: Ian Allan Ltd. ISBN 0-7110-1929-0

[20]   The Rail Regulator (ORR) is a statutory office, created with effect from 1 December 1993 by section 1 of the Railways Act 1993, for the independent economic regulation of the British railway industry. In 2015 ORR also took responsibilty to independently monitor and enforce the performance and efficiency of Highways England.

[21]   Two original minutes books are held in the Library Archives of The Institution of Civil Engineers in London.

[22]   https://www.rssb.co.uk/en accessed 07/09/2020.

[23]   Directive (EU) 2016/797 of the European Parliament and of the Council of 11 May 2016 on the interoperability of the rail system within the European Union (recast). Official Journal of the European Union, 26.05.2016. http://eur-lex.europa.eu/legal-content/EN/TXT/PDF/?uri=CELEX:32016L0797&from=EN, accessed on 11.01.2019.

[24]   Directive 2012/34/EU of the European Parliament and of the Council of 21 November 2012 establishing a single European railway area (recast). Official Journal of the European Union, 14.12.2012. http://eur-lex.europa.eu/legal-content/EN/TXT/ PDF/?uri=CELEX:32012L0034&from=EN, accessed on 11.01.2019.

[25]   Council Directive 96/48/EC of 23 July 1996 on the interoperability of the trans-European high-speed rail system. Official Journal of the European Communities, 17.09.1996. http://eur-lex.europa.eu/legal-content/EN/TXT/PDF/?uri=CELEX:319 96L0048&from=EN, accessed on 11.01.2019.

[26]   Directive 2001/16/EC of the European Parliament and of the Council of 19 March 2001 on the interoperability of the trans-European conventional rail system. Official Journal of the European Communities, 20.04.2001. http://eur-lex.eu- ropa.eu/legal-content/EN/TXT/PDF/?uri=CELEX:32001L0016&from=EN, accessed on 11.01.2019.

[27]   Diagrams based on Auer, F.: Infrastructure Management. PMC Media House GmbH, Leverkusen, 2017.

[28]   Commission Regulation (EU) No 1301/2014 of 18 November 2014 on the technical specifications for interoperability relating to the 'energy' subsystem of the rail system in the European Union. Official Journal of the European Union, 12.12.2014. http://eur-lex.europa.eu/legal-content/EN/TXT/PDF/?uri=CELEX:320 14R1301&from=EN, accessed on 11.01.2019.

[29]   Directive 2008/57/EC of the European Parliament and of the Council of 17 June 2008 of the rail system within the Community (recast). Official Journal of the European Union, 18.07.2008. http://eur-lex.europa.eu/ LexUriServ/LexUriServ.do ?uri=OJ:L:2008:191:0001:0045:EN:PDF, accessed on 11.01.2019.

[30]   Austrian Federal Ministry for Digital and Economic Affairs: Nationale/Europäische/ Internationale Normen. https://www.help.gv.at/Portal.Node/hlpd/public/content/256/ Seite.2560002.html, accessed on 19.07.2018.

[31]   Swiss Department for Environment, Transport, Energy and Communication: Gleisaushubrichtlinie: Planung von Gleisaushubarbeiten, Beurteilung und Entsorgung von Gleisaushub. Federal Transport Office, 22.08.2018. https://www. bav. admin.ch/dam/bav/de/dokumente/richtlinien/eisenbahn/gleisaushubrichtlinie. pdf. download.pdf/gleisaushubrichtlinie.pdf, accessed on 11.01.2019.

Bibliography – Chapter 2

[1]   EN 15273-2: Railway applications – Gauges – Part 2: Rolling stock gauge, July 2013. British Standards Institution.

[2]   EN 15273-1: Railway applications – Gauges – Part 1: General – Common rules for infrastructure and rolling stock, November 2017. British Standards Institution.

[3]   RSSB: Vehicle/Structures System Interface Committee Guide to British Gauging Practice (T926) 2013; RGS GIRT7073 Issue 2 Requirements for the Position of Infrastructure and for Defining and Maintaining Clearances

[4]   Rulebook 01.05: Entwerfen von Bahnanlagen – Streckenquerschnitte, 18.06.2015. ÖBB-Infrastruktur AG, Vienna.

[5]   Federal Republic of Germany: Eisenbahn-Bau- und Betriebsordnung (EBO). Federal Ministry of Justice and Consumer Protection, 08.05.1967, version of 26.07.2017. https://www.gesetze-im-internet.de/ebo/EBO.pdf, accessed on 11.01.2019.

[6]   Republic of Austria: Verordnung über den Bau und Betrieb von Eisenbahnen (Eisenbahnbau- und -betriebsverordnung – EisbBBV). Federal Ministry for Digital and Economic Affairs, 19.11.2008, version of 26.06.2014. https:// www.ris.bka.gv.at/GeltendeFassung.wxe?Abfrage=Bundesnormen&Gesetze snummer=20006077, accessed on 11.01.2019.

[7]   Swiss Department of Environment: Ausführungsbestimmungen zur Eisenbahn-verordnung (AB-EBV). Federal Office, 01.06.2016. https://www.bav.admin.ch/ bav/de/home/rechtliches/rechtsgrundlagen-vorschriften/ab-ebv.html, accessed on 16.07.2018.

[8]   Commission Regulation (EU) No 1299/2014 of 18 November 2014 on the technical specifications for interoperability relating to the 'infrastructure' subsystem of the rail system in the European Union. Official Journal of the European Union, 12.12.2014. http://eur-lex.europa.eu/legal-content/EN/TXT/PDF /?uri=CELEX:32014R1301&from=EN, accessed on 11.01.2019.

[9]   EN 15273-3: Railway applications – Gauges – Part 3: Structure Gauges, July 2013. British Standards Institution.

[10]   Freystein, H., Muncke, M., Schollmeier, P.: Handbuch Entwerfen von Bahnanlagen – Regelwerke, Planfeststellung, Bau, Betrieb, Instandhaltung. PMC Media House GmbH, Leverkusen, 2015.

[11]   UK standard structure gauge in Network Rail Standard NR/L3/TRK/2049/mod 7 "Gauging".

[12]   Directive 824.0101: Grundlagen für die Durchführung; Bautechnische Grundsätze bei Oberbauarbeiten, 01.01.2007. DB Netz AG, Munich.

[13]   Anweisung zur B51 „Oberbauvorschrift" und ZOV 55 „Lückenlose Gleise und verschweißte Weichen". Prävention gegen Gleisverwerfungen bzw. Tätigkeiten bei hohen Temperaturen. ÖBB-Infrastruktur AG, Vienna, 2017.

[14]   Arbeitsgruppe VöV (ed.): Handbuch Fahrbahnpraxis Normalspur. Bern, 2009.

[15]   Network Rail Standard NR/L2/TRK/3011 (2012). "Continuous Welded Rail (CWR) Track", Issue 7. Network Rail Standard NR/L3/TRK/002/E01 "Plain Line Tamping".

[16]   Directive 820.2010A04: Ausrüstungsstandard bei einer Gleisbelastung 10.000 Lt/d, 15.02.2018. DB Netz AG, Munich.

[17]   Directive 820.2010A05: Ausrüstungsstandard bei Gleisbelastung > 10.000 und < 30.000 Lt/d, 15.02.2018. DB Netz AG, Munich.

[18]   Directive 820.2010A06: Ausrüstungsstandard bei einer Gleisbelastung 30.000 Lt/d, 15.02.2018. DB Netz AG, Munich.

[19]   Rulebook 07.02.01: Planung und konstruktive Ausführung, 15.01.2015. ÖBB-Infrastruktur AG, Vienna.

[20]   Swiss Federal Council: Schweizerische Fahrdienstvorschriften (FDV). Federal Transport Office, 01.07.2016. https://www.bav.admin.ch/bav/de/home/rechtliches/rechtsgrundlagen-vorschriften/fdv-2016.html, accessed on 18.03.2019.

[21]   Network Rail Legal services freedom of information response 3.1.2014 from: https://www.whatdotheyknow.com/request/classification_of_lines, accessed on 2/11/20.

[22]   Network Rail Standard NR/L3/TRK/049/mod 1 "Guidance and Principles".

[23]   Haarmann, A.: Die Kleinbahnen. Salzwasser-Verlag, Paderborn, 2013 (Reprint of 1896 original).

[24]   Haban, F. X.: Theoretische und experimentelle Untersuchung an Spannbeton-schwellen. Doctoral thesis, Munich Technical University, Department of Traffic Route Engineering, Munich, 2016.

[25]   Strach, H.: Geschichte der Eisenbahnen der Oesterreichisch-Ungarischen Monarchie, Vol. 1. Forgotten Books, K. u. K. Hofbuchhandlung, Vienna, 2017.

[26]   The Railway – British track since 1804 by Andrew Dow, published by Pen and Sword books, Barnsley, England.

[27]   Haarmann, A.: Das Eisenbahn-Geleise. Verlag von Wilhelm Engelmann, Leipzig, 1891.   http://books.google.com/books?id=3gS1cp5q0fMC&oe=UTF-8, accessed on 01.02.2019.

[28]   Licensed image – Denby Tramway-wagonway (Outram's Railroad) near Little Eaton – Heritage Digital Ltd. Cromford Mills, Cromford Derbyshire, DE4 3RQ.

[29]   Photograph copyright: "Head of Steam – Darlington Railway Museum, England".

[30]   Richard Spoors/Plasser & Theurer

[31]   Führer, G.: Gleiskonstruktionen. Transpress Verlag für Verkehrswesen, Berlin, 1987.

[32]   Fengler, W.: Spurführung. Lecture notes, Dresden Technical University, Railway System Design, Dresden. https://tu-dresden.de/bu/verkehr/ibv/gvb/ressourcen/dateien/download_gvb/lehrmaterialien/Grundl_G/G-02.pdf?lang=de, accessed on 01.02.2019.

[33]   EN 13674-1: Railway applications – Track – Rail – Part 1: Vignole railway rails 46 kg/m and above, February 2011. British Standards Institution.

[34]   Network Rail Standard NR/L3/TRK/2049/mod04 "Components".

[35]   Network Rail legacy rail profiles from https://britishsteel.co.uk/media/40814/flatbottomed-rail-dimensions-and-properties.pdf, accessed 02/11/2020.

[36]   Fendrich, L. and Fengler, W.: Handbuch Eisenbahninfrastruktur. Springer, Berlin, 2013.

[37]   From https://www.railwaysarchive.co.uk/docsummary.php?docID=465, accessed 02/11/2020.

[38]   Photo: Fabian Hansmann

[39]   Directive 820.2110: Spurweite und Schienenneigung, 01.01.2000. DB Netz AG, Munich.

[40]   Network Rail Standard NR/L2/TRK/2102 "Track Construction".

[41]   EN 13848-1: Railway applications – Track – Track geometry quality – Part 1: Characterization of track geometry, April 2019. British Standards Institution.

[42]   Rulebook 07.04: Vermessung von Gleisen und Weichen, 02.12.2015. ÖBB-Infrastruktur AG, Vienna.

[43]   Network Rail Standard NR/L2/TRK/2102 "Track Construction".

[44]   Rulebook 01.03: Linienführung von Gleisen, 02.09.2016. ÖBB-Infrastruktur AG, Vienna.

[45]   PMC Rail International Academy GmbH: Rolling marks. https://www.trackopedia.info/encyclopedia/infrastructure/superstructure/rails/rolling-marks/, accessed on 09.10.2018.

[46]   Directive 820.2040A05: Führungen und Fangvorrichtungen, 01.04.2008. DB Netz AG, Munich.

[47]   Photo: Richard Spoors

[48]   PMC Rail International Academy GmbH: Infrastructure. https://www.trackopedia.info/encyclopedia/infrastructure/, accessed on 24.03.2019.

[49]   Hempe, T.: Life Cycle Management bei der Infrastruktur. Wo stehen wir bei der DB Netz AG? Presentation at the Concrete Sleeper Symposium. Berlin, 2017. http://www.betonschwellenindustrie.de/wp-content/uploads/Betonschwellenindustrie-Symposium-2017-K-Vortrag-8-Thomas-Hempe.pdf, accessed on 18.03.2019.

[50]   Network Rail Standard NR/L2/TRK/2102 "Track Construction".

[51]   Network Rail Standard NR/L2/TRK/2102 "Track Construction".

[52]   Commission Directive 2011/71/EU of 26 July 2011 amending Directive 98/8/EC of the European Parliament and of the Council to include creosote as an active substance in Annex I thereto. Official Journal of the European Union, 27.07.2011. https://eur-lex.europa.eu/legal-content/EN/TXT/PDF/?uri=CELEX:32011L0071&from=EN, accessed on 18.03.2019.

[53]   Directive 820.2010: Ausrüstungsstandard Schotteroberbau für Gleise und Weichen, 01.06.2013. DB Netz AG, Munich.

[54]   Deutsche Bahn Standard 918 144: Technische Lieferbedingungen Holzschwellen, 01.12.2007. DB Netz AG, Frankfurt.

[55]   Technical terms of delivery 07.09.07: Gleis- und Weichenschwellen aus Holz, 01.01.2017. ÖBB-Infrastruktur AG, Vienna.

[56]   Network Rail Standard NR/L2 /TRK/029 "Wood Sleepers, Bearers and Longitudinal Timbers".

[57]   DIN 5904:1995-11: Steel sleepers – Dimensions, static values, steel types. Beuth-Verlag, Berlin.

[58]   ISO 6305-3:1983-10-01: Railway components – technical delivery requirements – Part 3: Steel sleepers. International Organisation for Standardisation, Geneva.

[59]   Endmann, K.: Die Y-Stahlschwelle. Der Eisenbahningenieur (2002), Issue 11, pp. 18ff.

[60]   Standard drawing log 15.5000: Stahlschwelle St 82 K, 05.05.2005. DB Systemtechnik, Minden.

[61]   British Steel steel sleeper 436 from https://britishsteel.co.uk/media/282844/sleeper436.pdf, accessed 23/11/2020.

[62]   Standard drawing log 54.15.7001 NO: Stahlschwelle St 98 Y-No-650-54, 20.06.2003. DB Systemtechnik, Minden.

[63]   Standard drawing log 54.15.7000 NO: Stahlschwelle St 98 Y-No-600-54, 20.06.2003. DB Systemtechnik, Minden.

[64]   Monier, J.: Système de traverses et supports en ciment et fer applicables aux voies, chemins ferrés et non ferrés. French patent no. 120989, 1877.

[65]   Standard drawing 17208/10: Spannbetonschwelle L2, 14.06.2006. ÖBB-Infrastruktur AG, Vienna.

[66]   Standard drawing 17207/10: Spannbetonschwelle K1, 12.05.2005. ÖBB-Infrastruktur AG, Vienna.

[67]   Standard drawing 17211: Spannbetonschwelle HDS, 19.09.2016. ÖBB-Infrastruktur AG, Vienna.

[68]   DB Netz AG, Deutsche Bahn AG, Betonschwellenindustrie e. V. (eds.): Betonschwellen, Feste Fahrbahn, Fertigteiltragplatten, Komponenten im Netz der Deutschen Bahn AG. http://www.betonschwellenindustrie.de/wp-content/uploads/BSI-17010_Schwellenbroschuere_screen.pdf, accessed on 24.03.2019.

[69]   Vigier Rail AG: B 06 FS-Flachschwelle. http://www.vigier-rail.ch/fileadmin/media/vigier-rail/downloads/Dokumentationen/5.1.2_B_06_FS_Deutsch_08.2016.pdf, accessed on 24.03.2019.

[70]   Network Rail Standard NR/L3/TRK/2049/mod04 "Components".

[71]   Freudenstein, S., Iliev, D., Ahmad, N.: Die Kontaktspannung zwischen elastisch besohlten Schwellen und Schotter. Eisenbahntechnische Rundschau (2011), Issue 5, pp. 13ff.

[72]   Loy, H.: Körperschall-/Erschütterungsschutz durch besohlte Schwellen – Wirkung und Grenzen. Eisenbahntechnische Rundschau (2012), Issue 12, pp. 66ff.

[73]   Iliev, D.: Die horizontale Gleislagestabilität des Schotteroberbaus mit konventionellen und elastisch besohlten Schwellen. Doctoral thesis, Munich Technical University, Department of Traffic Route Engineering, Munich, 2011.

[74]   Photos: Richard Spoors

[75]   Diagram courtesy of Pandrol, Worksop, England

[76]   Marx, L. and Moßmann, D.: Arbeitsverfahren für die Instandhaltung des Oberbaus. Bahn Fachverlag, Berlin, 2011.

[77]   PMC Rail International Academy GmbH: Protection of the tracks. https://www.trackopedia.info/encyclopedia/infrastructure/superstructure/protection-of-the-tracks/, accessed on 31.10.2018.

[78]   Directive 820.0110: Oberbaubezeichnungen, 01.04.2008. DB Netz AG, Munich.

[79]   Rulebook 07.01.00.02: Kurzbezeichnungen am Oberbau, 24.08.2016. ÖBB-Infrastruktur AG, Vienna.

[80]   EN 13450: Aggregates for railway ballast, May 2013. British Standards Institution.

[81]   Achs, G.: Schotterflug bei hohen Geschwindigkeiten. Eisenbahntechnische Rundschau (2014), March (Austria Special), pp. 37f.

[82]   Directive 820.2010A08: Tieferkehren der Schwellenfächer in Gleisen, die mit v > 140 km/h befahren werden, 01.06.2013. DB Netz AG, Munich.

[83]   Directive 820.2010A07: Regelbettungsquerschnitte, 01.06.2013. DB Netz AG, Munich.

[84]   Network Rail Standard NR/L2/TRK/2102 "Track Construction".

[85]   Network Rail Standard NR/L2/TRK/2102 "Track Construction".

[86]   Network Rail Standard NR/L2/TRK/2102 "Track Construction".

[87]   Fischer, R., Göbel, C., Lieberenz, K. et al.: Handbuch Erdbauwerke der Bahnen – Planung, Bemessung, Ausführung, Instandhaltung. PMC Media House GmbH, Leverkusen, 2013.

[88]   Berg, G. and Henker, H.: Weichen. Transpress Verlag für Verkehrswesen, Berlin, 1978.

[89]   Pixabay. https://pixabay.com/de/.

[90]   Curr, J.: The coal viewer, and engine builder's practical companion. Printed for the author by John Northall, Sheffield, 1797. https://www.e-rara.ch/doi/10.3931/e-rara-16332, accessed on 18.03.2019.

[91]   Veit, P., Sanierung, Instandsetzung und Neubau im Lebenszyklus. Der Eisenbahningenieur (2011), September, pp. 26ff.

[92]   Photo: Richard Spoors

Bibliography – Chapter 3

[1]   Federal Republic of Germany: Eisenbahn-Bau- und Betriebsordnung (EBO). Federal Ministry of Justice and Consumer Protection, 08.05.1967, version of 26.07.2017. https://www.gesetze-im-internet.de/ebo/EBO.pdf, accessed on 11.01.2019.

[2]   Swiss Department of Environment: Ausführungsbestimmungen zur Eisenbahnverordnung (AB-EBV). Federal Office, 01.06.2016. https://www.bav.admin.ch/bav/ de/ home/rechtliches/rechtsgrundlagen-vorschriften/ab-ebv.html, accessed on 16.07.2018.

[3]   Republic of Austria: Verordnung über den Bau und Betrieb von Eisenbahnen (Eisenbahnbau- und -betriebsverordnung – EisbBBV). Federal Ministry for Digital and Economic Affairs, 19.11.2008, version of 26.06.2014. https:// www.ris. bka.gv.at/GeltendeFassung.wxe?Abfrage=Bundesnormen&Gesetzesnumm er=20006077, accessed on 11.01.2019.

[4]   Network Rail Standard NR/L3/TRK/2049/mod02 Track Design Manual "Mathematics" B.1.1 Curve Formulae.

[5] Pixabay. https://pixabay.com/de/.

[6] Schramm, G.: Der Gleisbogen. Otto Elsner Verlagsgesellschaft, Darmstadt, 1962.

[7] EN 13803: Railway applications – Track – Track alignment design parameters – Track gauges 1 435 mm and wider, June 2017. British Standards Institution.

[8] Directive 800.0110: Linienführung, 01.12.2015. DB Netz AG, Munich.

[9] Rulebook 01.03: Linienführung von Gleisen, 02.09.2016. ÖBB-Infrastruktur AG, Vienna.

[10] R I-22046: Geometrische Gestaltung der Fahrbahn für Normalspur, 13.01.2010. SBB Rulebook, Bern.

[11] Network Rail Standard NR/L3/TRK/2049/mod01 Track Design Manual "Guidance and Principles" A.6.2 Curving Design Values – Guidance on Circular Curves.

[12] Network Rail Standard NR/L3/TRK/2049/mod01 Track Design Manual "Guidance and Principles" A.6.2 Curving Design Values – Guidance on Circular Curves.

[13] Commission Regulation (EU) No 1299/2014 of 18 November 2014 on the technical specifications for interoperability relating to the 'infrastructure' subsystem of the rail system in the European Union. Official Journal of the European Union, 12.12.2014. http://eur-lex.europa.eu/legal-content/EN/TXT/PDF/?uri=CELEX:32014R1301&from=EN, accessed on 11.01.2019.

[14] Jochim, H. E. and Lademann, F.: Planung von Bahnanlagen. Grundlagen – Planung – Berechnung. Fachbuchverlag Leipzig im Carl Hanser Verlag, Munich, 2009.

[15] Arbeitsgruppe VöV (ed.): Handbuch Fahrbahnpraxis Normalspur. Bern, 2009.

[16] Rulebook 07.04: Vermessung von Gleisen und Weichen, 02.12.2015. ÖBB-Infrastruktur AG, Vienna.

[17] Network Rail Standard NR/L3/TRK/2049/mod01 Track Design Manual "Guidance and Principles" A.6.3: Curving Design Values – Guidance on Transitions.

[18] Fendrich, L. and Fengler, W.: Handbuch Eisenbahninfrastruktur. Springer, Berlin, 2013.

[19] Network Rail Standard NR/L2/TRK/2102 Issue 8 2017 "Design and Construction of Track".

[20] Rulebook 01.05: Entwerfen von Bahnanlagen – Streckenquerschnitte, 18.06.2015. ÖBB-Infrastruktur AG, Vienna.

[21] Network Rail Standard NR/L3/TRK/2049/mod01 "Guidance and Principles" A.6.3: Curving Design Values – Guidance on Transitions.

[22] Network Rail Standard NR/L3/TRK/2049/mod01 "Guidance and Principles" A.6.3: Curving Design Values – Guidance on Transitions.

[23] Network Rail Standard NR/L2/TRK/2102 Issue 8 2017 "Design and Construction of Track".

[24] Jochim, H. E. and Lademann, F.: Planung von Bahnanlagen. Grundlagen – Planung – Berechnung. Fachbuchverlag Leipzig im Carl Hanser Verlag, Munich, 2018.

[25] Network Rail Standard NR/L3/TRK/2049/mod01 Track Design Manual "Guidance and Principles" A.6.2: Curving Design Values – Guidance on Circular Curves and NR/L3/TRK/2049/mod02 "Mathematics" B.1.2: Compound Curves, Reverse Curves

[26] Railway Group Standard RGS GC/RT5021 issue 5 "Track System Requirements" – 2.2 General horizontal alignment requirements

[27] Network Rail Standard NR/L3/TRK/2049/mod01 Track Design Manual – "Guidance and Principles" A.6.2: Curving Design Values – Guidance on Circular Curves and NR/L3/TRK/2049/mod02 "Mathematics" B.1.2: Compound Curves, Reverse Curves

[28] Network Rail Standard NR/L3/TRK/2049/mod02 Track Design Manual "Mathematics" – B.5.6: Relative Lateral Movement (Buffer Locking Considerations).

[29]   Network Rail Standard NR/L3/TRK/2049/mod02 – Track Design Manual "Mathematics" – B.1.1: Curve Formulae.

[30]   Network Rail Standard NR/L2/TRK/2102 Issue 8 2017 "Design and Construction of Track" 7.2 Vertical alignment – new construction.

[31]   Network Rail Standard NR/L3/TRK/2049/mod02 Track Design Manual – "Guidance and Principles" – A.7.6 – Vertical Curves and Gradients.

[32]   Network Rail Standard NR/L3/TRK/2049/mod02 Track Design Manual – "Mathematics" – B.2.1: Vertical Curves Formulae.

Bibliography – Chapter 4

[1]   Pixabay. https://pixabay.com/de/.

[2]   EN 13848-1: Railway applications – Track – Track geometry quality – Part 1: Characterization of track geometry, April 2019. British Standards Institution.

[3]   Kipper, R. and Gerber, U.: Gleislagefehler – Ursachen, Messung und Bewertung. Eisenbahntechnische Rundschau (2013), Issue 12, pp. 40ff.

[4]   Network Rail Standard NR/L2/TRK/2102 "Design and Construction of Track" – Section 3 Definitions.

[5]   British Rail Civil Engineering Handbook No. 25 and British Railway Track 4th edition 1971 The Permanent Way Institution.

[6]   British Railway Track 6th edition 1993 The Permanent Way Institution and Railway Group Standard GC/EH0038 "Track Recording Handbook".

[7]   Network Rail Standard NR/L2/TRK/001/Mod 11 (2015) Inspection and maintenance of Permanent Way. "Track geometry – Inspections and minimum actions" Issue 8.

[8]   Management of Cyclic Top – a booklet published in 2002 by the (British) Railway Civil Engineering Conference.

[9]   Landgraf, M.: Der Einfluss von Unterbau, Untergrund und Wasserwegigkeit auf die Gleislagequalität. Master thesis, Graz Technical University, Institute of Railways and Transport, Graz, 2011.

[10]   EN 13848-5: Railway applications – Track – Track geometry quality - Part 5: Geometric quality levels – Plain line, switches and crossings, September 2017. British Standards Institution.

[11]   Photograph: Richard Spoors

[12]   RSSB https://www.rssb.co.uk/standards-and-safety/using-standards/Defining-the future-of-Standards/Applying-RSSB-Published-Standards-After-Brexit. Accessed 17/11/2020.

[13]   Network Rail Standard NR/L2/TRK/001/Mod 11 (Issue 8 2015) "Track geometry – Inspections and minimum actions".

[14]   EN 13848-3: Railway applications – Track – Track geometry quality – Part 3: Measuring systems – Track construction and maintenance machines, June 2009. British Standards Institution.

[15]   EN 13231-1: Railway applications – Track – Acceptance of works – Part 1: Works on ballasted track – Plain line, switches and crossings, May 2013. British Standards Institution.

[16]   Network Rail Standard NR/L2/TRK/3100 "Topographic, Engineering, Land and Measured Building Surveying" Strategy and General – Issue 5 and Mod 1 "Track".

[17]   Klotzinger, E.: Lehrbehelf – Gleis. Training CD, Plasser & Theurer, 1994.

[18]   Sinning, A. and Naumann, K.: Trimble – Vormessen mit Inertialmesssystem. Essay for publication in Gleisbauwelt, 2018.

[19]   Copyright and with kind permission, Abtus, Ltd. Suffolk, England.

[20]   British Railway Track 4th edition 1971 The Permanent Way Institution.

[21]   Network Rail Standard NR/L3/TRK/3241 – Issue 3 2020 – "Marking of Track for Tamping Machines".

[22]   Network Rail Standard NR/L3/TRK/002/E01 – Issue 1 2008 – "Plain Line Tamping".

Bibliography – Chapter 5

[1]   Hansmann, F.: Innovative Messdatenanalyse – Ein Beitrag für ein nachhaltiges Anlagenmanagement Gleis. Doctoral thesis, Graz Technical University, Institute of Railways and Transport, Graz, 2015.

[2]   EN 13306: Maintenance – Maintenance terminology, December 2017. British Standards Institution.

[3]   DIN 31051:2012-09: Grundlagen der Instandhaltung, September 2012. Beuth Verlag, Berlin.

[4]   Schenk, M. (ed.): Innovative Lösungen für die Instandhaltung von Anlagen. Tasking document, 11th Industrial Workshop "Co-operation in Plant Engineering", 17.06.2009, Magdeburg. Fraunhofer-Institut für Fabrikbetrieb und -automatisierung IFF, Fraunhofer Verlag, Stuttgart, 2009.

[5]   EN 13848-5: Railway applications – Track – Track geometry quality – Part 5: Geometric quality levels. Plain line, switches and crossings, September 2017. British Standards Institution.

[6]   Network Rail Track Standard NR/L2/TRK/001/mod11 – "Track geometry – Inspections and minimum actions".

[7]   Ekberg, A. and Paulsson, B. (eds.): INNOTRACK – Innovative Track Systems. Concluding technical report, Paris, 2010.

[8]   Hansmann, F.: Grundlagen zur Bettungsreinigung als Schlüssel. EIK – Eisenbahn Ingenieur Kompendium (2019), Issue 01, pp. 58ff.

[9]   EN 13231-1: Railway applications – Track – Acceptance of works – Part 1: Works on ballasted track – Plain line, switches and crossings, May 2013. British Standards Institution.

[10]   From https://www.harscorail.com/equipment/surfacing/stoneblower.html, accessed 02/12/2020.

[11]   Directive 824.3401: Kleine Instandsetzungsarbeiten am Oberbau durchführen; Stopf-Richtarbeiten mit Stopfgeräten oder mit Kleinstopfmaschinen durchführen, 01.01.2007. DB Netz AG, Munich.

[12]   Machine photographs copyright Matisa (UK) Ltd and Swietelsky (UK) Ltd.

[13]   Directive 824.3020: Gleise und Weichen durcharbeiten; Beseitigung von Geometrie-Einzelfehlern in Gleisen und Weichen mit Einzelfehlerstopfmaschinen, 01.01.2007. DB Netz AG, Munich.

[14]   Veit, P., Qualität im Gleis – Luxus oder Notwendigkeit? Der Eisenbahningenieur (2006), Issue 12, pp. 32ff.

[15]   Network Rail Track Standard NR/L2/TRK/2102 "Design and Construction of Track".

[16]   Directive 824.2200A01: Vorgaben für den Neu- bzw. Umbau, 01.06.2018. DB Netz AG, Munich.

[17]   Directive 824.2200A02: Vorgaben für die maschinelle Durcharbeitung, 01.06.2018. DB Netz AG, Munich.

[18]   Technical information: Stopfarbeiten im nicht verdichteten Schotterbett, 03.04.2015. ÖBB-Infrastruktur AG, Vienna.

[19]   Bd1351/1d: Abnahmetoleranzen bei Gleisbauarbeiten. 15.03.1999, SBB CFF FFS, Bern.

[20]   Network Rail Track Standards NR/L2/TRK/2102 "Design and Construction of Track". and NR/L2/TRK/001mod11 "Track geometry – Inspections and minimum actions".

[21]   Directive 824.2200: Maschinelle Stopf-Richtarbeiten durchführen, 01.06.2018. DB Netz AG, Munich.

[22]   Rulebook 07.04: Vermessung von Gleisen und Weichen, 02.12.2015. ÖBB-Infrastruktur AG, Vienna.

[23]   Network Rail Standard NR/L3/TRK/3241 "Marking of Track for Tamping Machines".

[24]   Technical note: Stopftiefeneinstellung bei Gleis- und Weichenstopfarbeiten, 02.05.2017. ÖBB-Infrastruktur AG, Vienna, 2017.

[25]   I-FW 001: Arbeitstechnische Qualifizierung von Gleisbaumaschinen, 01.05.2006. SBB Rulebook, Bern.

[26]   From https://www.sbrail.com/case-studies/sprinter-tamping/ and https://www.plassertheurer.com/en/spare-parts/machine-upgrades/smartalc.html, accessed 02/12/2020.

[27]   Photograph copyright Network Rail

[28]   Rießberger, K.: Schottergleise für hohe Geschwindigkeiten. Eisenbahntechnische Rundschau (2007), Issue 10, pp. 620ff.

Bibliography – Chapter 6

[1]   DIN 1961:2016-09: VOB Vergabe und Vertragsordnung für Bauleistungen. Beuth-Verlag, Berlin.

[2]   ÖNORM B 2110:2011 03 01: Allgemeine Vertragsbestimmungen für Bauleistungen. Österreichisches Normungsinstitut, Vienna.

[3]   EN 13231-1: Railway applications – Track – Acceptance of works – Part 1: Works on ballasted track – Plain line, switches and crossings, May 2013. British Standards Institution.

[4]   EN 13848-2: Railway applications – Track – Track geometry quality – Part 2: Measuring systems – Track recording vehicles, June 2006. British Standards Institution.

[5]   EN 13848-3: Railway applications – Track – Track geometry quality – Part 3: Measuring systems – Track construction and maintenance machines, June 2009. British Standards Institution.

[6]   EN 13848-4: Railway applications – Track – Track geometry quality – Part 4: Measuring systems – Manual and lightweight devices, January 2012. British Standards Institution.

[7]   EN 13848-1: Railway applications – Track – Track geometry quality – Part 1: Characterization of track geometry, April 2019. British Standards Institution.

[8]   EN 13848-5: Railway applications – Track – Track geometry quality – Part 5: Geometric quality levels – Plain line, switches and crossings, September 2017. British Standards Institution.

[9]   Network Rail Standards NR/L2/TRK/2102 "Design and Construction of Track".

[10]   Network Rail Track Engineering Form TEF3071 "OTM Site Check and Handback".

[11]   Network Rail Standard NR/L2/TRK/2102 "Design and Construction of Track".

[12]   Network Rail Standard NR/L2/TRK/001/mod12 "Track Geometry – Maintenance design requirements".

[13]   Railway Group Standard GC/RT5021 Issue 5 "Track System Requirements".

[14]   Network Rail Standard NR/L2/TRK/001/mod11 "Track Geometry – Inspections and minimum actions".

[15]   Network Rail Standard NR/L3/TRK/3261 "ATG (Absolute Track Geometry) Maintenance Process using ATG Geometry Methods".

Authors

Dipl.-Ing. Dr Fabian Hansmann

Fabian Hansmann graduated from Graz University of Technology in 2011 with a degree in civil engineering in the specialist field of environment and transportation. Between 2011 and 2016, he worked as a research associate at the Institute of Railway Engineering and Transport Economy at TU Graz, where he wrote his doctoral thesis (PhD) on "Digital asset management of 'track'". During his time at the Institute, he worked together with various railway infrastructure managers and industry partners on research projects on the topic of Life Cycle Costs and asset management. In 2015, he joined Plasser & Theurer in the Marketing and Communication department. His work has provided him with insights into the global developments in the field of track maintenance with a focus on the operation of large on-track machines. In 2019 he started working for Plasser American in Chesapeake (USA) as Senior Scientific Advisor for Railroad Engineering.

Wolfgang Nemetz

Wolfgang Nemetz completed his education at the Höhere Technische Bundes-Lehr- und Versuchsanstalt Wiener Neustadt in the field of electrical engineering. In 1985 he joined ÖBB, where he worked in track and switch surveying until 1991. Subsequently he acted as ÖBB operations manager of track/switch tamping machines and rail grinding trains. From 1998 to 2003 he took over the nationwide control of tamping machines and surveying. From 2003 to 2007, he was responsible for the technical and operational implementation of the entire mechanical track construction operation at ÖBB. After extensive organisational changes in the company, in 2007, he took over the management of the performance management department for the construction and maintenance of the Plant Service division of ÖBB-Infrastruktur BAU AG. From 2010 until his move to the private sector, he was most recently head of the Mechanical Engineering department at ÖBB-Infrastruktur AG. Since 2017, Wolfgang Nemetz has been working at the PMC Rail International Academy in project management, consulting and as Head of Division Development Austria.

Richard Spoors, MICE, FPWI

Richard Spoors (75) joined British Rail in 1964 as a Student Civil Engineer. After graduating from the University of Bradford with a degree in Civil Engineering, he honed his skills as a railway civil engineer on the Regions of British Rail, culminating with his appointment to the position of Regional Civil Engineer, Scotland, in 1989. The privatisation of the UK's railways took him into Railtrack and then Network Rail, from where he retired in 2005 having led the civil engineering team specifying and approving the upgrade of the West Coast Main Line. During the next period of Richard's career as a consultant he supported the UK Rail Regulator in their review of Network Rail's track policy. His particular interest in the management of all aspects of track saw him elected President of the Permanent Way Institution in 2004 and 2009.

List of abbreviations

A

AA	Vertical curve start
AB-EBV	Ausführungsbestimmungen zur Eisenbahnverordnung (Implementing Provisions of the Railway Regulation) (CH)
AE	Vertical curve end
AEG	Allgemeines Eisenbahngesetz (General Railways Act) (D)
AL	Alert Limit
AM	Vertical curve centre

B

BA	Curve start
Bar	Unit of pressure
BAV	Bundesamt für Verkehr (Federal Transport Office) (CH)
BE	Curve end
BEVVG	Bundeseisenbahnverkehrsverwaltungsgesetz (Federal Railway Transport Administrative Law) (D)
BHN	Unit for rail hardness (Brinell Hardness Number)
BMVIT	Bundesministerium für Verkehr, Innovation und Technologie (Federal Ministry for Transport, Innovation and Technology) (A)
BS	British Standard

C

CEN	European Committee for Standardisation CEN (Comité Européen de Normalisation)
CENELEC	European Committee for Electrotechnical Standardisation (Comité Européen de Normalisation Électrotechnique)
CWR	continuous welded rail

D

DACH	Acronym made up of the country codes of the German-speaking countries: Germany (D), Austria (A) and Switzerland (CH)
DAR	Data Acquisition Recorder
DB	Deutsche Bahn (German Railway)
DfT	Department for Transport
DTS	Dynamic Track Stabiliser
DRP	Data Recording Processor
DRS	Digital Recording System

E

EBA	Eisenbahn-Bundesamt (Federal Railway Authority) (D)
EBG	Eisenbahngesetz (Railways Act) (CH)

EBO	Eisenbahn-Bau- und Betriebsordnung (Railway Construction and Operations Regulation) (D)
EBV	Eisenbahnverordnung (Regulation on Railway Construction and Operation) (CH)
EFTA	European Free Trade Association
EC	European Community
EGMTPA	Equivalent Million Gross Tonnes per Annum
EisbBBV	Eisenbahnbau- und -betriebsverordnung (Railway Construction and Operations Regulation) (A)
EisbG	Eisenbahngesetz (Railways Act) (A)
EKAS	Eidgenössische Koordinationskommission für Arbeitssicherheit (Confederate Coordination Commission for Occupational Health and Safety) (CH)
ELTB	Eisenbahnspezifische Liste der Technischen Baubestimmungen (Railway-specific List of Technical Construction Regulations) (D)
EN	European Standard
ETSI	European Telecommunications Standards Institute (European Telecommunication Standards Institute)
EU	European Union
EW	Single switch
G	
T.G.R.T/d	Total gross register tons per day
GKP	Track checkpoint
GPS	Global Positioning System
GTG	Good track geometry
GV	Freight traffic
GVP	Track marking point
H	
HMRI	Her Majesty's Railway Inspectorate
HP	High point
HSTRC	High speed track recording coach
Hz	Unit of frequency (Hertz)
I	
IAL	Immediate Action Limit
IL	Intervention Limit
IMU	Inertial Measurement Unit
ISO	International Organisation for Standardisation
ITU	International Telecommunication Union
K	
km/h	Kilometres per hour

L
Lt/d · Load tonnes per day

M
MCRS · Multi-Channel Recording System
mph · miles per hour

N
NR · Network Rail

O
ÖBB · Österreichische Bundesbahnen (Austrian Federal Railways)
OHLE · Overhead line equipment
ORR · Office of Rail and Road

P
PV · Passenger traffic

R
R · Backward
RA · Ramp start
RCF · Rolling contact fatigue
RE · Ramp end
REA · Railway Engineers' Association
RIC · Agreement governing the exchange and use of coaches in international traffic (Regolamento Internazionale delle Carrozze) (CH)
RIL · Deutsche Bahn guideline
RIV · Agreement governing the exchange and use of freight wagons in international traffic (Regolamento Internazionale dei Veicoli) (CH)
R RTE · Supplementary provisions on sovereign regulations and technical standards in the RTE (CH)
RSSB · Railway Safety and Standards Board
RTE · Regelwerk Technik Eisenbahn (Railway Technical Rules and Standards) (CH)
RW · Railway Technical Rules and Standards of ÖBB

S
SBB · Schweizerische Bundesbahnen (Swiss Federal Railways)
SD · Standard deviation
SERA · Single European Railway Area
SFT · Stress free temperature
SIA · Swiss Association of Engineers and Architects
SN · Swiss standard

T

TC	Technical Committee of CEN
TEN	Trans-European Networks
TM	Technical Note (on ELTB guidelines) (D)
TRV	Track recording vehicle
TSI	Technical Specification for Interoperability
TSI LOC&PAS	Technical Specification for Interoperability – Locomotives and passenger rolling stock

U

ÜA	Transition curve start
ÜE	Transition curve end
UIC	International Union of Railways (Union Internationale des Chemins de fer)
UK	United Kingdom
USA	United States of America

V

V	Forward
VDV	Association of German Transport Undertakings
VöV	Association of Public Transport (CH)

Index

Advertisements